Nuclear Medicine

Radioactivity for Diagnosis and Therapy

Richard ZIMMERMANN

Illustrations by Pascal COUCHOT

17, avenue du Hoggar
Parc d'activités de Courtabœuf, BP 112
91944 Les Ulis Cedex A, France

Édition originale : *La Médecine nucléaire. La radioactivité au service du diagnostic et de la thérapie.* Richard Zimmermann. © EDP Sciences 2006

Traduction : Objectif traduction

Coordination de la traduction : Susan Brown

Cover illustrations:
PET (© Philips Medical Systems) ; heart (© GE/Amersham Healthcare) ; radiopharmaceutical preparation and imaging (© CIS bio international).

Imprimé en France

ISBN : 978-2-86883-962-6

Tous droits de traduction, d'adaptation et de reproduction par tous procédés, réservés pour tous pays. La loi du 11 mars 1957 n'autorisant, aux termes des alinéas 2 et 3 de l'article 41, d'une part, que les « copies ou reproductions strictement réservées à l'usage privé du copiste et non destinés à une utilisation collective », et d'autre part, que les analyses et les courtes citations dans un but d'exemple et d'illustration, « toute représentation intégrale, ou partielle, faite sans le consentement de l'auteur ou de ses ayants droits ou ayants cause est illicite » (alinéa 1er de l'article 40). Cette représentation ou reproduction, par quelque procédé que ce soit, constituerait donc une contrefaçon sanctionnée par les articles 425 et suivants du code pénal.

© EDP Sciences 2007

In memory of my father

This work is dedicated to all the anonymous persons who are directly or indirectly involved in the preparation, the handling and the application of radiopharmaceutical drugs. They are technicians, cyclotronists, engineers, pharmacists, biologists, clinicians, scientists, archivists, salesmen, specialists of radioprotection, safety, quality, logistics, environment, regulatory affairs, maintenance, etc., and without their precious contribution nuclear physicians would not be able to bring to their patients – who are often affected by extremely invalidating diseases, sometimes considered as incurable – the benefit of these little known, extremely complex and particularly efficient drugs.

Preface

Richard Zimmermann firmly believes in the huge possibilities offered by nuclear medicine. At a time when the public hears about Positron Emission Tomography, at a time when this new technology seems to give a second breath to our speciality, it was clever to propose a didactic work allowing to better understand nuclear medicine. Richard Zimmermann provides the reader with a book that is complete, describing with great accuracy yet in an attractive and comprehensive style all the technical, methodological and pharmaceutical aspects of nuclear medicine, covering also its diagnostic and therapeutic applications. This book proves to be a valuable information source for our young students as well as for hospital intern specialists. Others will find it very useful when promoting this too little known discipline. This work must be made available in all our university libraries, and it must be read by all those in charge of health politics.

Professor Patrick BOURGUET, MD, PhD
Professor of the University
Director of the Regional Centre for Cancer Prevention
Eugene Marquis Hospital, Rennes, France
Treasurer of the European Association of Nuclear Medicine

Introduction and Definitions

Nuclear medicine covers the area of a medical practice based on the resources of physics, its tools and products – nuclear meaning related to the nucleus of the atom – in order to be applied both for diagnosis and therapeutics. In both cases, a substance containing a radioactive isotope or **radionuclide** (one speaks about a labelled substance) is administered to a patient. It goes straight towards a biological tissue or organ after having selectively searched for it. The concentration of this radionuclide in the targeted tissue or organ is favoured by the design of the organic or biological substrate or **vector** on which it is grafted. The emitted radioactivity will then be used either to locate the radionuclide (diagnosis) or to initiate the deterioration of surrounding cells (therapy).

The term **radiotracer** refers to the notion of minute (trace) amounts of the substances in use, and also to the advantageous ability to "trace" the dissemination of the molecule in the body. The selection of the radionuclide, based on the nature of the emitted radiation, its physical properties, *i.e.* energy and half-life, and its chemical properties, will define the final purpose of this molecule, called a **radiopharmaceutical**. The diagnosis imaging technology, also called scintigraphy, is obtained with substances labelled with γ-emitter isotopes. The development of the imaging acquisition technology associated with powerful information technology software resulted in the development of the tomography technology, which involves cross sections and tri-dimensional images. Some radioactive elements can be used for therapeutic purposes thanks to their physico-chemical properties, based on the destructive effect of the ionising radiation emitted by the labelled substance. The use of these vectors in association with therapy radionuclides, essentially β^- or α emitters, is called **vectorised** or **metabolic radiotherapy**.

The use of radionuclides as an external source of radioactivity, of temporary radioactive implants for therapy purposes or of particle beam generators (neutrontherapy and protontherapy, which are both external therapies) is controlled by radiotherapists. Hence, it does little or not concern nuclear physicians. The same applies to **brachytherapy**, or **curietherapy** as it is known in some countries, a technology that is devoted to the use of permanent or temporary radioactive implants. These areas will however be described in this work. Finally, analogue sources (mainly X-rays) are used in **radiology** to obtain organ image data from different angles around the body.

CHAPTER 1

Nuclear Medicine, what for?

After about fifty years of practice and experience, nuclear medicine reached a turning point. The new imaging modalities that appeared on the market at the dawn of the new millennium, as well as the new molecules and therapeutic technologies associated to radioactivity open new and promising perspectives that fascinate experts from other medical disciplines, and more particularly oncologists, haematologists and neurologists.

This work does not intend to put forward new therapies and original answers to pathologies that seem hopeless. Physicians do have all the competencies required to prescribe the most appropriate treatments for specific patients and diseases. This book simply aims to provide detailed information, using a vocabulary that is as accessible as possible, to a public that is usually not even aware that this discipline does exist and that it brings a new breath of life to diagnosis and therapy, especially in oncology.

For a long time, therapy by means of nuclear medicine was restricted to very difficult cases and last chance treatments. Physicians usually think of using metabolic radiotherapy only after chemotherapy and external radiotherapy protocols repeatedly failed. One forgets much too quickly that Iodine 131 is almost always used in thyroid cancer treatment. For the past fifty years more than 90% of all thyroid cancers have been successfully treated, and definitely cured, thanks to this nuclear medicine method. However, one has to admit that, so far, this therapeutic success was the only one recorded following the use of this technology until very recently, by the end of the 90's.

Until then, the role of nuclear medicine was essentially limited to being a support for diagnosis, via every type of scintigraphy method developed to this day.

In this introductory chapter, we will see how patients can benefit from the knowledge acquired in nuclear medicine during the past half century, and learn about the revolutionary aspect of these new techniques and medicinal products; then, we will take a look at all the opportunities brought by this technology when it is associated to other innovative medical modalities. All these aspects will be developed in detail further in this book.

What do we call Cancer?

All living beings originate from a single cell which divided, grew and continued to multiply while it underwent differentiation, in order to form the specific cells of the various organs that make up an individual. The production of these cells follows a complex predefined process at a rhythm that is also predetermined. However mature bodies do not expand anymore, therefore the production of new cells concentrates on specific growth mechanisms as for hair or blood, and also as part of repair mechanisms such as skin regeneration or wound self-repair. As the lifespan of a cell is limited, its renewal is also necessary.

Taking into account the impressive number of cells which are required to build a complete body, the process regularly deviates, giving birth to cells with an unexpected structure. Also, cells are continually subject to external stress, called the toxic effect, and this factor also interferes with the process of cell reproduction. Although the body is well adapted to automatically correct or destroy these aberrant cells, it sometimes happens that these new entities find a more hospitable, or in any case less hostile, terrain in which they can reproduce identically. When these new types of cells are not rejected by the organism, they create a new tissue that is called a tumour.

Tumours may either be benign or malignant. Benign tumours are not cancers, because they do not propagate themselves. They are easily removed without consequence and without recurrence, and above all they do not represent a vital risk to the patient.

NUCLEAR MEDICINE, WHAT FOR?

Malignant tumours, on the other hand, are composed of abnormal cells which divide and grow in a wild fashion, invading the tissue to the point of destroying it or preventing its proper functioning. To the detriment of healthy tissue, they divide, spread and travel along the blood or lymphatic system, re-implanting themselves some distance away. The new colony formed is called a metastasis, but in fact it has the same properties as the original tumoral cells. In turn, these metastases colonise other tissues, causing the disease to spread further again.

Each tumoral cell is a malformation of a healthy cell of a very specific type. Therefore, it can be identified by the organ from which it originated. As the formation of a metastasis is only a remote colony, that is to say a relocated reproduction of these same cells, metastases of an identified type will display the same properties as the cells belonging to the original tumour. Thus, the primary tumour and the metastases of a tumour originating in the prostate will both be treated in an identical way, even if the latter are located in another organ far from the prostate. It is therefore important to determine the origin of a cancer, *i.e.* the primary tumour, in order to be able to treat the metastases, even long after the original tumour has been removed. It follows that one would not say of a person being treated for lung cancer and presenting with metastases of the liver, that they are suffering from liver cancer, but rather from a lung cancer that has spread. This person will be treated for lung cancer, a therapeutic protocol that greatly differs from liver cancer treatment.

Lymphomas and leukaemias are particular cancers that form in the blood precursor cells (hematopoietic system). These abnormal cells circulate in the blood and lymphatic systems and reproduce to the detriment of the production of normal blood cells. They are sometimes called liquid cancers in order to be distinguished from solid tumour cancers.

I. The Case of Thyroid Cancer

Iodine 131 plays a key role and so we will start with it. The earliest imaging trials, followed by the first therapeutic treatments of hyperthyroid disease with injected radioactivity, began in 1942. In 1946 it was demonstrated that not only did thyroid tumours disappear following Iodine 131 treatment, but also all metastases, thus proving the power of this technique. This incontestable efficacy is linked to the fact that thyroid tissue is the only tissue capable of absorbing iodine. This fixation includes also the metastases, as these tissues are originating from the thyroid cells. Today, this method remains essential for the diagnosis of thyroid diseases as well as for their treatment (*see below Chapter VI, Section I*). Unfortunately it remains the unique example of human tissue fixing a radionuclide in such a specific manner, and so therapeutic nuclear medicine remains unsatisfied with this unique but major success.

Nevertheless, iodine having demonstrated some physico-chemical advantages, it remained for a long time a privileged tool for nuclear medicine. On this basis, several other radioisotopes were used for the labelling of molecules for diagnosis purposes.

II. The Diagnosis Aspect

Nuclear medicine imaging is first of all a functional imaging tool: it allows to check if a tissue or an organ works, *i.e.* is alive. Contrarily to all other imaging modalities, nuclear medicine is the only one able to prove brain death for example. Magnetic Resonance Imaging (MRI), X-rays (X) or Ultrasound (US) are unable to make the difference between dead and living tissues, and will only provide a nice three-dimensional image of the brain. It is obvious that this imaging technology is not used in this extreme case as an electroencephalogram (EEG) will provide the same information in a more simple way. However this example shows that this technology is extremely powerful as it can be used to monitor the functioning of the brain, the heart (necrosis, infarction) or the growth rate of a tumour invading a tissue (*Chapter IV*). Therefore almost every organ can be visualised, and tracers are now available for nearly all tissues (bone, liver, kidney, heart, lung, gastro-oesophageal tract, etc.), and fluids (blood, cerebro-spinal liquid, urinary excretion tract, etc.).

The discovery of the utility of Thallium 201 in heart imaging, followed by that of several Technetium 99m derivatives and linked with the progress of the image acquisition technology, made this an essential cardiology tool. Nowadays, cardiology care units very frequently resort to nuclear medicine, and are its main users: almost all persons affected from an infarct undergo scintigraphy. These tools give an accurate cardiac pump function check.

Actually, the word scintigraphy stands for all of these two-dimensional imaging techniques (*Chapter IV, Section II*). Cross-section images can be obtained by associating a rotating camera and a powerful computerised calculation system, called tomoscintigraphy. This technique, which brought a new dimension to the technology, evolved in such a way that today three-dimensional imaging acquisition is possible. However, the amount of data to be analysed having simultaneously increased, medical applications had to wait until the end of the 90's, *i.e.* until the new calculator revolution, before they could take place in a realistic time frame with low cost computers. Three-dimensional imaging was in fact limited by computer power.

The main pathologies that have benefited from these imaging methods are:

– imaging of the lung with determination of the zones that can be reached by the inhaled air, and in parallel by the blood that will collect the oxygen in these alveolar areas (pulmonary embolism);

– bone scintigraphy, which allows to determine the metastatic zones and the development stage of a cancer;

– kidney scintigraphy, which allows to check if all renal filtration mechanisms are functioning properly (renal dysfunction);

– imaging of inflamed or infected tissues (in case of internal lesions, polyarthritis, appendicitis, etc.);

– and of course, localisation of all tumours and metastases which usually require a different molecule per type of tissue.

A non-exhaustive list of available products is provided in *Chapter IV* with details on their use.

In parallel to computer development, a new technology, the Positron Emission Tomography (PET), was introduced first in North America, and then in Europe. By the end of the 90's, the USA were the first to be properly equipped, whereas in Europe only Germany and Belgium had an adequate equipment network.

The introduction of this new technology was slower in France, Spain and Italy until 2001. At that time, some industries took the risk to install the specific and expensive manufacturing equipment while some governments arranged for dedicated cameras to be installed in public hospitals. Today, in 2006, a few other countries (*e.g.* Great Britain, Canada) are in a decision-making process concerning investment, but are still in a waiting phase and therefore remain under-equipped. North Africa and South America are slowly implementing the installation of cameras. PET imaging technology can only be made available in a country when manufacturing sites for the powerful diagnostic drug fludeoxyglucose (or FDG) are also made available there (*Chapter V*). In the meantime, this astonishing product has proven its efficacy as it offers unquestionable advantages:

– FDG is polyvalent. Its mechanism of action enables it to be integrated in any functioning or growing cell: in the brain and the heart of course, but also in tumours and metastases, which grow faster than the other surrounding "healthy" cells;

– almost all cancer types can benefit from this technique and even some small metastases can be detected;

– FDG is easy to use. The radioactivity completely disappears in less than 24 hours thanks to the short half-life (less than 2 hours) of the associated radionuclide (Fluorine 18);

– patients are reassured by the fact the involved radioactivity has limited concentration levels, and this in turn enables physicians to use this tool when monitoring the efficacy of a given treatment;

– finally it seems that non-experts can interpret the images themselves. This is not true however, as false negative as well as false-positive results do exist also, but the image remains reassuring for the physician himself.

PET technology, together with its FDG tracer, is recognised as being an extremely useful diagnosis modality for evaluating tumours: head and neck tumours (particularly tongue cancers), pulmonary nodules, gastro-oesophageal cancers, differentiation between pancreas chronic inflammation and cancer, colorectal cancers, ovarian cancers, detection of bone marrow cancer metastases, melanomas, Hodgkin disease and non-Hodgkin lymphoma. Disease extension (staging), chemotherapy or radiotherapy treatment response level, and actual possibilities of surgery can also be evaluated. Benign tumours can often be differentiated from malignant ones in the absence of any response. This modality could be used to evaluate breast cancer, but other available techniques can provide equivalent information at lower cost. In the latter case, FDG remains an interesting tool to estimate the disease's level of extension, and even to monitor patients at risk of relapsing. However, this technique is less interesting for the diagnosis of renal or prostate cancers, for which more efficient imaging technologies are available.

It must be reminded that there is a difference between this non exhaustive list of indications and the official list approved by the authorities as part of the FDG Marketing Authorisation (*Chapter V, Section V*). Important efforts have been made by clinicians to demonstrate that the not yet approved indications are valid at the level of a larger population and their integration in the official list could take place within the next few years. In reality, the use of PET technology differs from one continent to another, from one country to another and even from one centre to another.

The latest technological revolution associated computer science with PET, thus resulting in the development of mixed tools. PET/CT cameras, which combine a three-dimensional PET detection system with an X-ray tomography equipment can generate images in which the distribution of the FDG tracer can be superimposed with a three-dimensional view of the body. The localisation of the tumour becomes much more precise, to such an extent that, for example, surgeons can better outline the tumour excision area, hence improving its removal.

First and foremost, the association of PET and FDG is dedicated to oncology. Nevertheless FDG can be useful in analysing some brain functions (definition of the affected areas or brain damages following a brain stroke, evolution of neuro-degenerative diseases) or cardiac functions (viability of the cardiac tissue following a heart stroke). Nowadays, due to the limited access to this technology, it is only rarely used for these indications.

Nuclear Medicine

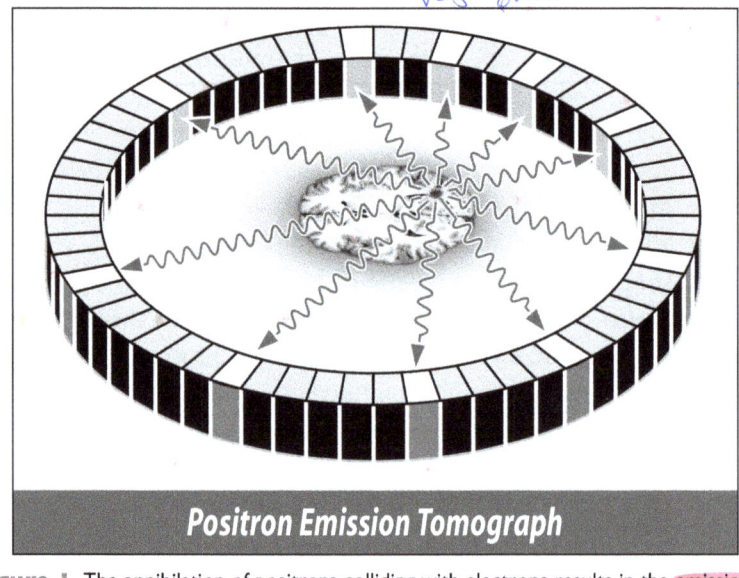

Positron Emission Tomograph

Figure 1. The annihilation of positrons colliding with electrons results in the emission of gamma photons. Those rays are taken into account by detectors placed on a level plane; this allows the acquisition of a section image of the radioactivity source (the scheme shows the section of a brain with a tumour). The image of the whole body is reconstructed by superimposing hundreds of such cross sections.

On the other hand, PET is a highly valuable tool for brain function study, and the development of new positron emitter labelled tracers, others than FDG and specific to neurological mechanisms, will probably result in improved diagnosis accuracy for neurodegenerative diseases such as Alzheimer's, Parkinson's or Huntington's diseases. The development, in parallel, of MRI technology becomes a necessity for patients affected from these diseases.

On the neurology side, a few new substances have already been marketed, but they are only used in very difficult cases. As of today, all of these diagnosis drugs are based on the SPECT technology involving gamma emitters. In fact, a whole new research field opened up due to population ageing.

III. The Therapeutic Aspect

Apart from some rheumatology affections, the therapy is mainly devoted to oncology in a large manner (including haematology).

1 Cancer Therapy

Beta-minus, beta-plus and alpha rays induce cellular destruction that can be turned to our advantage to destroy unwanted cells. On the contrary, radioactivity cannot be of any help concerning pathologies for which cells have to be stabilised, and even less so when these cells have to be regenerated. Only abnormal or supernumerary (excrescence) cells (tumours) are targeted.

Tumours can in fact be destroyed under the effect of a powerful external beam of radioactivity (RX, α, β^-, γ, neutron or proton), but this technique is part of the external radiotherapy, which is the domain of the radiotherapist, not of the nuclear physician.

Internal or metabolic radiotherapy, which is part of the therapeutic nuclear medicine, consists in injecting in a patient a radioactive substance that will be integrated in the cells to be destroyed by radiations (*Chapter VI*). Iodine 131, which is described in the introduction as being useful in thyroid cancer therapy, is probably the best example of this process. In recent years, new molecules appeared that proved their real efficacy concerning some very specific pathologies, efficacy which can be observed in the treatment of patients affected by non-Hodgkin lymphoma and resisting to classical therapies (*Chapter VI, Section III*). The treatments for less common tumoral pathologies (pheochromocytomes, neuroblastomes, polycythaemie, thrombocythaemie, etc…) and for chronic lymphocyte and myelocyte leukaemia had been known for a while, but the number of treated patients was directly linked to the very low incidence of these diseases. Some cases of non-transplantable liver cancers also benefited from a particular radiotherapy protocol in care centres well-known for their specificity. Finally, it was demonstrated a few years ago that in the absence of a total recovery, specific radiolabelled substances could significantly reduce the pain caused by bone metastases (*Chapter VI, Section II*).

Today, these therapies are either restricted to patients in classical therapeutic failure mode, or to very limited and well identified subgroups of patients. Metabolic radiotherapy still needs to demonstrate its efficiency on a larger scale and in first line. In this regard, clinical trials are currently underway and already show encouraging results.

Other new molecules also undergoing clinical trials will become available in hospitals during the few next years, particularly for the therapy of lung or colon cancers, lymphoma, myeloma and leukaemia.

Great progress was also made in surgery techniques, breast cancer benefiting from the most innovating and effective one, the sentinel lymph node detection (*Chapter IV, Section III*). If this technique was implemented for all breast cancer patients, it should result in a dramatic decrease of cancer recurrence. Moreover, this technique is a lot less traumatic than the surgical lymphatic system ablation which is the current procedure. This technique could also be adapted to melanoma therapy as well.

Treating Cancers

Today, several cancer treatment methods are available to physicians, and they are becoming increasingly effective. In the last ten years, oncology has reached a decisive turning point, and undergone its own revolution. All issues have not been solved yet, but great steps have been made for the benefit of patients. There exists a typical, efficient and well-defined therapy protocol for each cancer, once its development stage has been correctly evaluated.

It is important, for the well-being of patients, that they should know which therapy will be applied to them, and what it involves. It is even more important that they should be warned of potential side effects, in order to anticipate and, when possible, alleviate them. If the slightest doubt persists, patients are free to get a second opinion from another specialist. That specialist generally finds a sufficient base in the first analysis and evaluation to give his own opinion.

Therapeutic methods are fairly numerous, demonstrating the complexity of care, but also the lack of a universal treatment. Let's take a quick look at the current available treatments.

Surgery

The initial idea of physically removing a tumour is also the oldest method. A tumoral mass can be removed extremely cleanly with surgery. In order to ensure that the tumour has been completely removed, as any tumoral cell residue could re-colonise the area, surgeons must also take out some of the surrounding healthy tissue. If a tumour is well localised in uniform tissue, its removal (excision) can be carried out in an extremely clean manner. This is the case with melanomas, or surface tumours, but unfortunately the operation involved will leave a large scar as it usually necessitates the removal of a relatively wide and deep area of the skin. In spite of the unsightly scarring, this type of situation is ideal, however it does not occur very frequently. A growing tumour can spread to the surrounding tissue, making surgery more delicate. A melanoma on the foot, in direct contact with tendons or bone, is more difficult to treat than if it was on the thigh. A prostate tumour that extends beyond the prostate and touches the bladder raises other problems. Therefore, the decision to operate depends on the size of the tumour, its location, the patient's general health condition and the method of surgery required, including the type of anaesthetic. A surgeon can only remove what he can see, that is to say, the primary tumour, and any visible metastases if they are not numerous. When the disease has developed too far or has spread, another therapy has to be envisaged, or a complementary therapy must be integrated in the protocol.

Side effects are the same as for traditional surgery, and are linked to the consequences of the anaesthesia, as well as to the fatigue generated by the healing process.

External Radiotherapy

A certain analogy can be drawn between the use of radiation to destroy a tumour and surgery in the sense that the beam of radiation replaces the scalpel. In the same way, intense cold or a laser can be used to burn tissue. Nevertheless, it is not a question of removing the tumour, but of destroying the malignant cells on-site.

In the external radiotherapy technique, a beam of high energy beta or gamma radiation is directed against the tumoral mass. The treatment is applied to both surface tumours and internal tumours. The beam is not selective and destroys many cells as it passes over them, including healthy cells. In order to overcome this problem, a deep tumour is bombarded from various angles; therefore, it receives a cumulative dose whereas healthy cells only receive a fraction of the radiation. The radioactive source revolves around the patient in a predefined path, depending on the size and shape of the tumour. The tools used have become so precise that the technique is applied when surgery becomes too invasive, or else for delicate procedures (*e.g.* for brain tumours). Treatment is spread over several sessions that are spaced out in time. The procedure is not painful, but patients may feel tired. The irradiated skin may be affected (change of colour and irritation) and hair in this area may fall out. A combination of surgery and radiotherapy is sometimes suggested. The cells are then subject to intense radiation designed to destroy them, and the surgeon removes the tumoral mass to prevent any surviving cells from developing once more.

A new variable high energy protontherapy technology (beam of protons) is now being implemented in some countries. This very expensive equipment has the major advantage of limiting the energy deposition in the tumour alone with an accuracy to within 1 mm, thus avoiding damage to many surrounding healthy cells. Its use is primarily intended for eye, brain as well as children cancers.

Internal Radiotherapy

When the option of accessing certain cavities and introducing a source of radiation into them is available, this method, also called internal radiotherapy or curietherapy, is preferred. Cervix cancer is one of internal radiotherapy's target diseases. A radioactive implant is temporarily placed in contact with the tumour. Obviously, the radioactivity completely disappears as soon as the implant is removed. Brachytherapy, an extension of curietherapy, uses implants which are permanently placed in the tissue.

Chemotherapy

This well-known technique consists of using toxic substances to destroy the undesirable cells. Considerable progress has been made in this field, the main difficulty being to find molecules that destroy the tumoral mass selectively without affecting healthy cells. Today these drugs are like any others, and the term chemotherapy has become inappropriate; indeed, it could be applied to many drugs that treat by destroying, like antibiotics, because it is about using a toxic chemical substance to destroy a cell.

One substance on its own does not do all the work. In order to ensure that the treatment is effective, with a number of mechanisms acting at tumoral cell level, it is preferable to give patients several active substances (a cocktail); these substances must be taken in a specific order, in doses that are pre-established and of proven effectiveness (the protocol), over several months. It is important that this protocol is followed scrupulously, because it takes into account the evolution of the tumoral mass over the course of the treatment. It is a race against the disease which involves a chemical substance that must destroy the cells faster than they can reproduce, while at the same time avoiding damage to surrounding healthy cells.

Most chemotherapy is based on cell renewal mechanisms. The aim is to prevent the tumoral mass from developing by blocking its reproductive and cell division system, while at the same time destroying existing cells. Tumoral cells reproduce faster than normal cells; this is why they are able to spread more quickly, and to the detriment of healthy cells. As the reproductive mechanisms of tumoral cells are identical to those of healthy cells, some healthy cells that reproduce as quickly as the tumoral cells are also affected. Thus, hair which grows regularly finds its growth blocked. The same goes for cells lining the stomach and intestines. These mechanisms explain hair loss and digestive disorders.

Therefore, a complementary treatment is prescribed along with the chemotherapy; in particular, substances are given to calm nausea and limit vomiting. Chemotherapy also affects the reproduction of blood cells (red and white corpuscles, platelets), which must be compensated for by transfusions. Once treatment is over, patients are given substances to accelerate the regeneration of these blood cells.

All chemotherapy is exhausting and often demoralizing. An optimistic environment is more than necessary.

Hormone therapy is a special type of chemotherapy based on blocking tumoral cells' access to certain hormones they need for growth. The method either acts on the capture mechanism of these substances, or on their production mechanism, sometimes going as far as eliminating the source of these hormones (ablation of the testicles or the ovaries).

In immunotherapy, the body's immune system activity is encouraged with elements designed to naturally fight the invasion of foreign cells. Interferon and the interleukins are products which have already shown advantages in cancer therapy.

Metabolic Radiotherapy and Radioimmunotherapy

Rather than trying to destroy cancerous cells using the toxicity of a substance, the question very soon arises of finding out if it might also be appropriate to use the destructive properties of radioactivity. A radioisotope is attached to a molecule, or vector, capable of recognising cancerous cells in order to target and then destroy these cells selectively. The product, injected intravenously, must be capable of concentrating in the tumoral cells wherever these may be. This is the principle behind using therapy radiopharmaceuticals which take advantage of one of the metabolic mechanisms of cancerous cells or of the immune system, becoming involved in the functioning of this mechanism. The therapeutic part of this book takes a closer look at this principle.

Generally speaking, as with a number of new products in the course of evaluation, patients may be offered a treatment that has not yet totally proven to be effective but from which a certain degree of improvement is expected. Patients are guaranteed to benefit from the advantages of the traditional protocol, that is to say they can expect results in their health at least equivalent to those expected if they had not agreed to take part in this clinical study. All patients are free to accept or refuse to participate in a clinical study; should they agree, they must sign a document indicating their "informed" agreement.

2 Another Therapeutic Application: Rheumatology

The major part of therapeutic nuclear medicine is devoted to cancer therapy. Even if the patient is initially looked after by a physician who is a specialist of the affected organ (gastro-enterologist for a stomach cancer, hepatologist for the liver, urologist for the prostate), the therapeutic follow up will eventually take place in hospital under the responsibility of an oncologist (or haematologist).

Among all the other slowly evolving diseases that could be taken into account by nuclear medicine, only rheumatology found some practical applications with this technique. Rheumatoid polyarthritis remains the best example of this, as solutions were found independently of the type of affected joints, simply by using different radionuclides with similar properties: this technology is called radiosynoviorthesis (*Chapter VI, Section II*). This treatment is currently limited to particular and extreme cases as it is very costly.

Finally, let us not forget that today only one product labelled with an alpha-emitter can be found on the market. It is only sold in a few countries, for the treatment of ankylosing spondylarthritis (also called Bechterev disease) (*Chapter VI, Section V*).

IV. Other Aspects in this Area

In order to understand nuclear medicine as a diagnosis modality available for all human organs and tissues, and also to appreciate its therapeutic benefits in probably every cancer pathology, it seemed that delving deeper into the principles of this not so well-known science and taking a closer look at its constraints, regulations, technologies, costs, limitations and hopes would be beneficial to all of us. In order to facilitate reading, most of the technical and scientific words are explained in details in a glossary at the end of this book.

The first chapter provides some basic knowledge on the different types of radiation used in nuclear medicine, and also on the risks linked to radioactivity (*Chapter III*), then we take a look at the techniques and products used for diagnosis (*Chapter IV*) as well as for therapy (*Chapter VI*). Taking into account the interesting evolution of Positron Emission Tomography, a full chapter is dedicated to this technology (*Chapter V*). Also, as a lot of hope is placed on new drugs, it seemed for us to be of interest to describe the entire development process of a radiopharmaceutical, and more generally, of a drug (*Chapter VII*). The application of these products, starting from the manufacturing tool up to the injection to a patient, is detailed in *Chapter VIII*. Finally, we will take a look into the future, far beyond the products currently under development (*Chapter IX*).

Let us start first with some historical background in order to relate the origins of this (very) young science that is nuclear medicine.

CHAPTER II

A Little Bit of History...

Nuclear medicine is not itself a vanguard technology, as its first applications appeared only a few years following the discovery of radioactivity at the end of the nineteenth century. The development of this science benefited from the parallel evolution of three distinct technologies: nuclear physics beginning with the discovery of radioactivity right up until the development of radionuclides with well adapted half-lives and energy for clinical applications, radionuclide chemistry, *i.e.* chemistry allowing the incorporation of radioactive atoms into organic molecules, and eventually instrumentation with, in particular, the development of detection equipment and associated computers and software.

Nuclear medicine is an extension of the discovery of radioactivity that first allowed radiology imaging technique development. In 1895, Henri Becquerel (1852-1908; 1903 Physics Nobel Prize winner) made the chance discovery that certain substances, such as uranium salts, blackened photographic plates in the absence of light; at the same time, he observed that placing some objects between this source (subsequently called radioactive) and this plate could reduce the intensity of the radiation. In parallel, and in fact during the same year, Wilhelm Conrad Roentgen (1845-1923; 1901 Physics Nobel Prize winner) developed equipment that could generate an unknown radiation, unknown because indefinable and original, which he called X. X-rays show similar properties to the radiation discovered by Henri Becquerel, but it would be a few years before scientists understood the true link between both of them.

However, less than one year later, Wilhelm Roentgen succeeded in creating the first image of an X-rayed hand and demonstrated that human tissues behave differently depending on their density. This first radiography of Mrs Roentgen's hand, dating from 1896, became famous and opened the way for a new discipline of medicine, radiology. The recognition of the value and advantages of this technology was such that special radiology services were created within five years, and military medicine services in particular developed mobile imaging services.

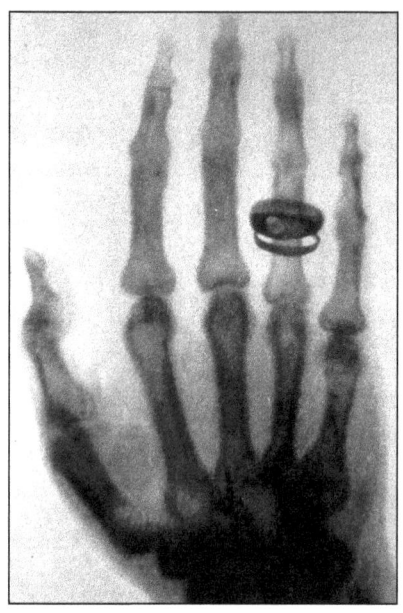

Figure 2. Radiography of Mrs Roentgen's hand, taken in 1896.

The atom model, with its nucleus made of protons and neutrons around which electrons are gravitating, was proposed by Ernest Rutherford (1871-1937; 1908 Chemistry Nobel Prize winner) in 1911. This model was improved by Niels Bohr (1885-1962; 1922 Physics Nobel Prize winner) in 1913. The concept of isotope was introduced by Frederick Soddy (1877-1956; 1921 Chemistry Nobel Prize winner) only in 1913.

The first natural radioactive substances, radium and polonium, were isolated in 1898 by Pierre Curie (1859-1906; 1903 Physics Nobel Prize winner) and his wife Marie Sklodowska Curie (1867-1934; 1903 Physics Nobel Prize winner and 1911 Chemistry Nobel Prize winner). Following this discovery, numerous experiences were performed with uranium and radium. As early as 1900, German-born physicists Otto Walkoff and Friedrich Giesel observed the effect of radium on the skin, which was similar to that seen a few years earlier with X-rays. Out of pure scientific curiosity, Pierre Curie repeated the experience by placing a radium source directly in contact with his skin during approximately 10 hours. The resulting red blotch transformed into a wound that would take 50 days to heal. Henri Becquerel accidentally observed the same result on his chest after having carried a sealed radium source in his jacket pocket during several hours. French physicians Henri Alexandre Danlos (1844-1912) and Eugène Bloch made use of these substances as early as 1901 to treat cutaneous tuberculous affections. In 1903, American-born Graham Bell (1847-1922) suggested placing radioactive sources on tumours. This marked the beginning of Curietherapy, which would be developed later thanks to the work of Claudius Regaud (1870-1940) on radiosensitivity. This scientist laid down the basis of modern radiotherapy (use of selective radiation, multiplication of the irradiation focuses and areas, distribution of the radiation in space and time, calculation of optimal doses).

Precise radiation measurements were made possible with the discovery of special counters by Hans Wilhelm Geiger (1885-1945) in 1908-1910. The system was improved by Walter Müller in 1928 with the development of counting tubes which allowed quantification of these radiations. Later, these counters would be replaced by scintillation crystals associated to photomultipliers (Samuel Crowe Curran and William Baker 1944, Hartmut Kallman 1947, Robert Hofstadter 1948).

From then on, Georg Charles de Hevesy (1885-1966; 1943 Chemistry Nobel Prize winner) could follow the diffusion of radioactive substances in solutions, then in plants (1911). In 1913, Frederick Proescher published a first study on the distribution of radium injected intravenously with a therapeutic aim. In 1924, Georg de Hevesy, J. A. Christiansen and Sven Lomholt used Lead 210 and Bismuth 210 on animals. On this basis, in 1926, Herman Blumgart (1895-1977), an American physician, injected himself a few millicuries of Bismuth 214 in order to follow his own blood circulation, and with the help of Saul Weiss repeated the procedure with other volunteers and patients.

A vast area of research first saw the light of dawn in 1934, when the production of artificial radionuclides was discovered by Frédéric Joliot (1900-1958) and his wife Irène Curie (1897-1956). The couple were awarded the Chemistry Nobel Prize in 1935 with this discovery, without which nuclear medicine would have quickly ground to a halt. Most of the natural radioisotopes, the only ones known at that time, were inadequate for medical applications due to their nuclear characteristics (radiation and half-life). As early as 1937, John Livingood, Fred Fairbrother and Glenn Seaborg (1912-1999; 1951 Physics Nobel Prize winner) discovered Iron 59. One year later, John Livingood and Glenn Seaborg developed the production of Iodine 131 and Cobalt 60. Today, all three isotopes are still in use in nuclear medicine and radiotherapy.

Technetium 99, of which the metastable nuclide 99m is the most important artificial isotope in imaging, was discovered in 1937 by Emilio Segrè (1905-1989; 1959 Physics Nobel Prize winner for the discovery of the antiproton), Carlo Perrier and Glenn Seaborg.

In 1936, John Lawrence, brother of the cyclotron inventor Ernest Orlando Lawrence (1901-1958; 1939 Physics Nobel Prize winner), injected for the first time a radioactive substance to a patient, namely Phosphorus 32, as a treatment for leukaemia.

In 1938, Saul Hertz, Arthur Roberts and Robley Evans performed the first research studies on the thyroid with Iodine 131; then, for the first time and as early as 1942, they treated patients affected with hyperthyroidia. In 1946, on the same basis, Samuel Seidlin (1895-1955), Leonidas Marinelli and Eleanor Oshry demonstrated that it was possible, following treatment with Iodine 131, to destroy all the metastases in a patient suffering from thyroid cancer. Later, Benedict Cassen (1902-1972) used radioactive iodine to demonstrate that thyroidal nodules accumulate iodine, thus allowing benign and malignant tumours to be differentiated from one another. These results had a major impact on the development of nuclear medicine as they demonstrated without the shadow of a doubt that this technique had great potential. Today it remains the most efficient method for thyroid cancer treatment. In 1951, Iodine 131, in the form of sodium iodide, became the first radiopharmaceutical to be approved by the Food and Drug Administration (FDA).

It was not until 1959 that radioimmunoanalysis methods, which allow the quantification of tiny amounts of substances present in serums, were developed by Rosalyn Sussman Yalow (born in 1921, 1977 Physiology and Medicine Nobel Prize winner) and Solomon Berson (1919-1972). This technology provided the basis for the *in vitro* biological analysis based on radionuclides.

As for imaging methods, one had to wait until 1950, when the linear scanner was invented by Benedict Cassen (1902-1972), to see a significant evolution. From that moment on, a sum of radioactive values measured around a body area could be transformed into an image using this tool. In 1953, Gordon Brownell and William Sweet built the first detector to allow the coincidence counting of radiation emitted by positron annihilation. Hal Anger (1920-2005) developed in 1957 the gamma camera (scintillation camera), capable of measuring the radioactivity of a surface in one go, rather than point by point as with the linear scanner. This technology has considerably improved since, in terms of detector sensitivity and resolution, as well as in terms of measurement speed. As a result, the most recent equipment is able to provide computerised superimposable three-dimensional colour images. Tomography, the two dimensional imaging technique also called tomodensimetry, appeared in 1962 thanks to David Kuhl. The evolution of this technique resulted in the invention of Tomoscintigraphy or Single Photon Emission Computed Tomography and Positron Emission Tomography. At the same time, this technology was integrated to X-ray scanning tools: the result was the creation of X-ray Tomodensimetry or X-scan Computed Tomography. Hal Anger contributed to the development of other medical imaging tools, inventing the tomographic scanner, the full body scanner and the positron camera. In fact John Keyes was at the origin of the development of the first SPET camera (1976) and Ronald Jaszczak developed the first dedicated SPET camera. Positron emitters applications in medical imaging are mostly based on the research work of Michel Ter-Pogossian (1925-1996) and Michael Phelps (1939-).

Technetium 99m only found its first real medicinal application when the first commercial generators were made available by Louis Stang Jr. and Powell (Jim) Richards from the Brookhaven National Laboratory in 1960. The first positron generator (Rubidium 82) was approved by the Food and Drug Administration (FDA) in 1989.

In 1963, Henri Wagner obtained the very first lung images using radiolabelled albumin aggregates, and in 1973 William Strauss introduced stress testing for cardiac imaging. A few years later, in 1978, David Goldenberg developed the first radiolabelled antibodies for tumour imaging but the FDA did not approve this new product as being a medicinal drug until 1992.

Finally, the PET technology would never have reached its current stage of development without the improvement of the cyclotron technology, and overall without the discovery of the FDG (fludeoxyglucose). This molecule was synthesised for the first time by the team of Al Wolf (1923-1998) and Joanna Fowler (1942-) from the Brookhaven National Laboratory in 1976, on the basis of an idea by Lou Sokoloff and Mark Reivich who had already worked with Carbon 14 labelled glucose. The first image of a patient injected with FDG was obtained by the team of Michael Phelps, Henry Huang, Edward Hoffman and David Kuhl at the University of Pennsylvania.

The term "Nuclear medicine" itself was coined in the 50's by the physician Marshall Brucer (1913-1994), who was actively engaged, along with the US Army, in favour of the use of radionuclides in medicine. As a teacher, he created a study programme in the US allowing physicians to get a degree as well as the very first authorisations to handle radionuclides within the medical field.

CHAPTER III

Some Basic Notions of Radiation

Natural radioactivity first appeared as the earth was created, and the radionuclides that can still be found in soil today are in fact long period isotopes or decay nuclides dating from the original matter. As a consequence, man has always been subjected to environmental radioactivity, whether it be from terrestrial or cosmic origin, and of course we continue to find these radionuclides in our food and indirectly in every one of our body cells. It is essential to give more details on the nature, the origin, the amount and the effects of this type of radiation, as radioactivity is associated with risk and danger.

In particular, when discussing radioactivity hazards one must take into consideration both the levels of natural and artificial radioactivity to which man is subjected daily. Therefore, let us start by explaining some of the words that are recurrent throughout this book, and defining units of measurement before taking a closer look at their effect on cells.

A **radioisotope** is an unstable atom that becomes another entity after a while, stable or also unstable, while at the same time emitting energy in form of **particles or radiation**. The origin of this energy emission is called **radioactive source**. Radioactivity is linked to the **decay phenomenon**, in other words the reduction over time of the amount of emitted radiation. The **period** (or **half-life**) determines the length of time required before half of the existing amount of material turns into another isotope. The period is a constant for a given radioisotope and can cover durations from a fraction of a second to several billion years. As a result of decay being an inverted exponential function of time, the activity of a source reduces to one tenth after only slightly more than three periods, and to one thousandth of its initial activity (2^{10} = 1,024) after only ten periods.

The word **radioisotope** should only be used for elements with the same chemical entity (Iodine 123, Iodine 124, Iodine 131 etc…), while the plural word **radionuclides** refers to every radioactive element from the periodical table. To simplify matters, the word "hot" may be used to describe a radioactive substance or isotope, as opposed to "cold" for a stable isotope.

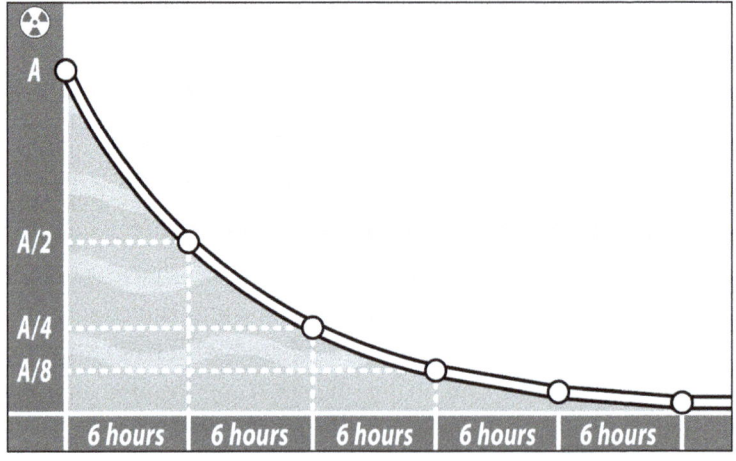

Figure 3. Decay and half-life: a radioactive substance looses half of its activity (A) during each period (or half-life) following a very regular decay curve. The decay of Technetium 99m with 6 hours half-life is given as an example.

Besides its half-life, two other parameters define a radionuclide: its type of emitted radiation and the level of energy associated with this radiation. These three characteristics, type of radiation, energy and half-life, single out the radionuclides that are relevant to nuclear medicine among the thousands of known ones. The associated chemical and biological properties are decisive criteria of selection, independent of radioactivity.

I. Different Types of Radiation

Four different types of radiation in particular have found an application in nuclear medicine, but other different types of radiation, offering increased therapy potential, are beginning to emerge.

The **gamma (γ) ray** corresponds to the emission of short wavelength and variable energy photons. This radiation translates the loss of excessive energy in the nucleus and its transformation into a more stable state, as opposed to the production of X-rays, obtained after excitation and ionisation of electrons. Gamma rays are ideal diagnosis tools. As they are very penetrating, they can go through large thicknesses of matter and travel for hundreds of meters in the air. Dense materials such as lead, wolfram or uranium, very thick concrete or deep water are able to stop them or, at least, strongly attenuate them. The emitted radiation differs for each isotope, thus allowing to better identify the originating radionuclide.

Beta-plus (β^+) ray emit positively loaded electrons called positrons. Positrons result from the transformation of a supernumerary proton into a neutron, neutrino and positron, the latter being nothing else than an anti-electron. The weightless neutrino does not play a role in nuclear medicine. The mass of the resulting isotope remains unchanged, but its atomic number is reduced by one unit. The positive electrons emitted by the radioisotopes in the continuous energy spectrum will eventually collide with a negative electron along their ejection path. The collision will lead to the annihilation of the matter and to a transformation into pure energy, in the form of two 511 keV photons that have the particularity to travel in two exactly opposite directions. If two detectors are placed on each side of the emission source, it becomes possible to calculate the precise origin of that source after several collisions. This method would be very precise, if it wasn't for the fact that the resulting image represents the distribution of the collision sites, and not the positron emission sites. Some high energy positrons can travel for several millimetres before encountering an electron, and then only producing their two photons. The analysis of the journey of all these photons forms the basis of the imaging technology, called **Positron Emission Tomography** (PET).

Beta-minus rays (β^-) are made of electrons, namely particles of identical mass to that of the positron, which are negatively loaded and travel at high speed. This radiation is the result of the transformation of an excessive neutron into a proton, electron and antineutrino. The weightless antineutrino, like the neutrino, does not play a role in nuclear medicine. The proton participates to the reorganisation of the nucleus and to the transformation of the initial radionuclide into a new element with an atomic number increased by one unit. Only electrons are ejected in a continuous energy spectrum which depends directly on the radioisotope. These electrons can go through several centimetres thick walls and travel several metres up in the air, but, most of the time, they are quickly absorbed by matter where they can sometimes generate X-rays while their excess energy is transformed into heat. They also generate free radicals that can conduct to molecular rearrangements; thus, they present a high destruction potential, and some radioisotopes selected for their specific half-life and energy can be used for localized cell destruction. Therefore, beta-minus emitters are efficient when used in therapy, mainly in oncology.

The ionising property of an isotope used in therapy is evaluated in the average penetrating distance that is directly linked to its energy. These values are comprised between a few millimetres and a few centimetres, thus allowing choice among several isotopes in order to find the most appropriate one to treat or destroy a tumour.

It is obvious that the higher the energy, the longer the penetrating distance, and, as a consequence, the higher the risk for healthy cells will be. At the same time, environmental irradiation risk levels increase.

A nucleus that decays when emitting beta rays can leave the resulting radioisotope in an excited state, which in turn will immediately emit gamma rays to return to its stable stage. In some cases this excited state is itself stable for a well-defined and measurable period. Known as the **metastable** state, it is identified with the letter "m" beside the isotope's mass number. The best known example of this is Technetium 99m, with a half-life of 6.01 hours. Technetium 99m is a pure γ emitter obtained after decay and β⁻ emission of Molybdenum 99 (half-life 8.04 days).

The **alpha ray** (α) corresponds to the spontaneous generation of a heavy particle consisting in a naked nucleus formed by two neutrons and two protons, which is in fact the nucleus of helium. This entity being 7,000 times heavier than the electron (*i.e.* β⁻ radiation), it is stopped by a very small amount of matter. Only a few centimetres of air absorb the radiation, and a simple sheet of paper is enough protection against alpha rays.

The alpha particle is stopped by organic tissue. It ionises the molecule and indirectly cuts it or allows it to be transformed chemically. If the molecule in question is a vital part of the reproduction mechanism (DNA or RNA for example), this interaction will usually result in the death of the cell.

X-rays (RX), correspond to the emission of photons (light) of a particular type, as a consequence of the excitation of electrons. It was the first radiation to be observed and artificially produced, and it is mostly known as an external body imaging source. It gave birth to radiology, which developed considerably further due to improved detector equipment and later, computer technology. X-rays can also result as secondary radiation from the interaction between a high-energy beta minus ray and the matter in which it is absorbed. This secondary radiation, also called *Bremsstrahlung*, originates from the impact area between the electron and the matter, *i.e.* in the mass. If this reaction takes place in the protective elements, it can happen that the residual thickness left after the generation of the X-ray is not sufficient anymore, thus generating supplementary radiation protection issues.

X-rays or γ rays can be generated as a consequence of the **electron capture** phenomenon. If the nucleus does not have sufficient energy to emit a positron, the proton in excess can be transformed into a neutron by trapping an electron which gravitates too closely to the nucleus. The resulting empty space will itself be completed by taking another electron from a higher layer (orbital) and by emitting in parallel either X-rays or γ rays.

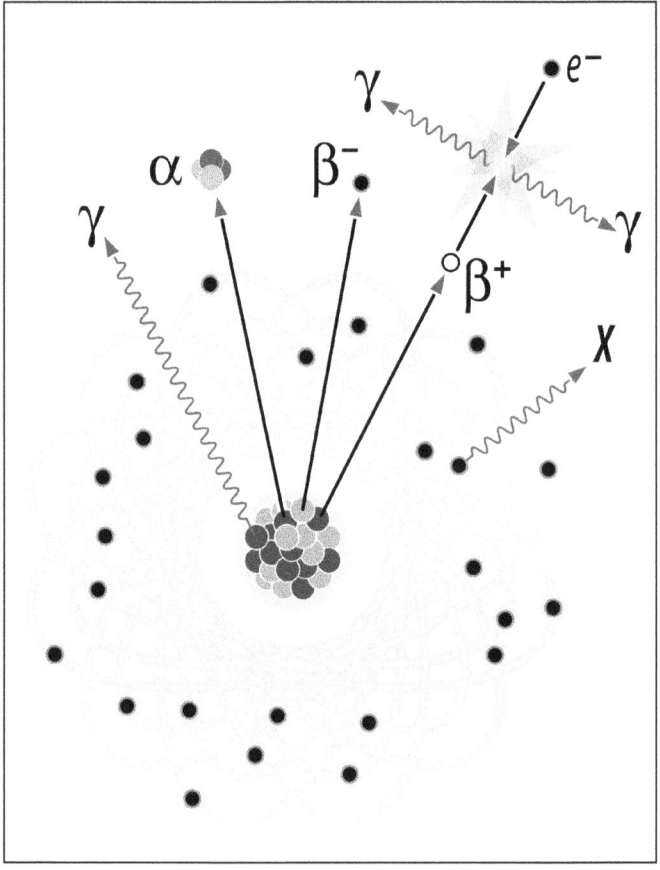

Figure 4. The different types of radioactive radiations used in nuclear medicine (gamma, beta-plus, beta-minus, alpha and X-ray).

If the excess energy of a radio-transformation or **internal conversion** is transferred to another layer of electrons, rather than via γ rays emissions, it can happen that an electron is ejected. These electrons, called **Auger electrons**, are much less energetic than β⁻. However, they show a therapeutic application potential analogue to that of alpha emitters, as their sphere of activity is limited to a few tenths of a millimetre and as a result are only useful for the destruction of a few layers of cells.

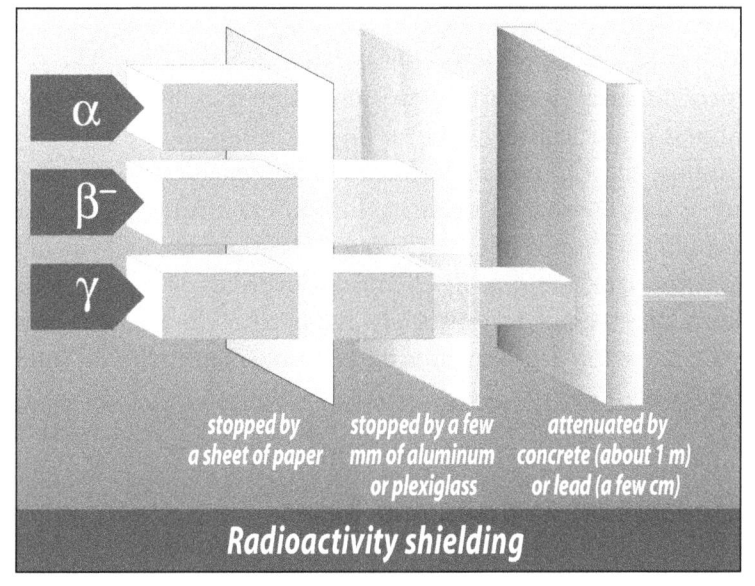

Figure 5. Radioactivity shielding: the different radiations are stopped in different ways, depending on the shielding material and its density.

Neutrons (n) and protons (p) are also used within the framework of external radiotherapy techniques. These particles are generated with specific tools and their use represents a large part of the external radiotherapy technology, out of the scope of nuclear medicine.

Finally let us not forget that we, as human beings, are under the influence of cosmic rays from outer space (in particular muons, but also γ, n and X-rays) on a daily basis, and that, should the necessity arise, these rays could only be stopped by several tens of centimetres of lead.

In fact, only very few radioisotopes are emitters of a single type of radiation. Technetium 99m is an almost pure γ emitter, Fluorine 18 is a pure β⁺ emitter, and Yttrium 90 is a pure β⁻ emitter. Most of the other radionuclides are emitters of at least two radiation types, most frequently β⁻ and γ or X-rays. If the γ part is not too high, the radionuclide will be found helpful for both diagnosis and therapy. If the β⁻ part associated to X-ray is sufficiently low, this isotope could be taken into consideration for imaging. The other physico-chemical properties of this element will determine its usefulness in nuclear medicine.

Among the thousands of known radionuclides, more than a hundred are potentially interesting for nuclear medicine use. Taking in account the cost of manufacturing and supply, the half-life of the radioisotope and the specific chemistry involved, only a few tens of radioisotopes remain as realistically usable. Therefore, let us concentrate on these few radionuclides and demonstrate their real interest through examples.

II. Measurement Units and Doses

The activity of a defined amount of radioactive nuclide is measured in **becquerels** (Bq). One becquerel corresponds to the disintegration of one atom per second (the becquerel is a measurement unit that replaced the **curie** (Ci), 1 curie being equivalent to 37 billions of becquerels). Taking into account the fact that the radioactivity level emitted by these substances is usually high, unit multiples are more frequently used, and particularly megabecquerels (MBq, 10^6 Bq) gigabecquerels (GBq, 10^9 Bq) and terabecquerels (TBq, 10^{12} Bq). Doses injected to a patient for an imaging procedure usually range from a few hundred to a few thousand MBq.

The emitted radiation transfers a certain quantity of its energy to the material or the tissue by absorption. This amount of energy is expressed in **grays** (Gy), one gray being equivalent to one joule per kilogram. Grays measure the **absorbed dose** and replace the former **rad** units (1 Gy = 100 rads). It is possible to link the emitted radiation to the absorbed one, but the equation must take into account the energy of the source, the distance to the source, the duration of irradiation and the nature of the radiation. This calculation is much simplified in the case of a fully absorbed source, as almost all of the radiation will affect surrounding tissues as long as the radiation has not decayed, or if the radioisotope is not biologically eliminated.

As it is mandatory to take into account the nature of the radiation and its effects on tissue, the absorbed dose will be corrected by a weighting factor in order to obtain an **equivalent dose**, expressed in **sieverts** (Sv). Sieverts replace the former **rem** units (1 Sv = 100 rems).

Absorbed Dose, Equivalent Dose and Effective Dose

The effect of radiation on a given organism depends on the dose absorbed into the tissues, but also on the type of radiation and the sensitivity of the tissues or organs irradiated.

The dose absorbed into the tissue or organ (**absorbed dose**), expressed in grays (Gy) corresponds to the quantity of energy imparted per unit mass of matter. A gray corresponds to one joule per kilogram.

The effect on the tissue will be different, however, depending on whether the particle is energetic or not. Photons generate other effects in the tissues than neutrons or alpha radiation. The absorbed dose is therefore corrected by a weighting factor which allows for the **equivalent dose** to be obtained. Thus X, γ or β radiation are similar in their effects and do not need to be corrected (weighting factor equal to 1), while neutrons are assigned a weighting factor between 5 and 20 depending on their level of energy. The highest coefficient is applied to neutrons as their energy is between 100 and 2,000 keV, and reduces again after that. This is explained by the fact that very high energy neutrons pass through tissue so rapidly that they do less damage than if they passed at a lower energy level. Extremely ionising α radiation is in turn assigned a weighting factor of 20.

Tissue sensitivity must also be taken into consideration, because each organ reacts differently to radiation. From this is deduced the **effective dose** which is quite simply the sum of the equivalent doses corrected by a weighting factor according to the tissue irradiated. The most sensitive tissues are quite obviously the gonads (weighting factor of 0.2), followed by bone marrow, the colon, the lungs and the stomach (0.12). A factor of 0.05 is assigned to the bladder, the breasts, the liver, the oesophagus and the thyroid, and 0.01 to the skin and the bone surface. At an equivalent dose, the effect on the testicles would be twelve times greater than on the skin.

These coefficients enable the calculation of external and internal exposure dose limit values for people working with nuclear substances. Effective dose limits are currently set at 20 mSv in Europe, but new texts in the course of preparation will lower these values still further.

Man ingests on average 100 Bq of Carbon 14 per day (half-life 5,730 years) and almost as much Potassium 40 (half life 1.3 billions of years). The latter accumulates in bones and alone accounts for about 6,000 Bq in a 75kg adult. On the other hand, due to the constant equilibrium preservation of the different salts present in body cells (homeostasis phenomenon), potassium does not accumulate anymore in the human adult body and its content remains therefore stable. As a consequence, the average radioactivity of an adult human body stays around 150 Bq/kg, which brings the radioactivity of an adult around 8,000 to 10,000 Bq.

In order to provide information on the effect of natural radioactivity on man, it is easier to compare equivalent doses using sievert units. External radiation as received by man, of terrestrial origin and caused by non-ingested elements such as soil radioactivity, accounts for about 0.4 millisieverts (mSv) over one year. The ingested part, which irradiates the body one hundred percent, participates to 0.3 mSv among which 0.18 are solely due to Potassium 40. Man also inhales radioactive gases, in particular radon and its daughter radionuclides, as they emanate from naturally radioactive material such as granite, present in concrete, in the walls of our homes etc. These gases are a cause of supplementary internal irradiation and represent in total between 1 and 20 mSv over a year, depending on location. A figure of 1.3 mSv is usually taken as an average value.

Cosmic radiation contributes in average at 0.4 mSv for our yearly irradiation, but this value highly depends on altitude. It can vary from 0.3 mSv at sea level up to 2.0 mSv at 5,000 meters altitude. Doses absorbed by passengers during a flight, and in the extreme by astronauts, are even higher.

The sum of all these individual values reaches approximately 2.4 mSv, which corresponds to the average radioactivity dose as absorbed by a European or American individual living in an area with a sedimentary soil only containing little amounts of radioactive substance. In the following paragraphs, we will use this value as reference when comparing the different doses from artificial (medical) origins to which patients are subjected. It is obvious that this figure must be adapted depending on whether one lives on the

SOME BASIC NOTIONS OF RADIATION

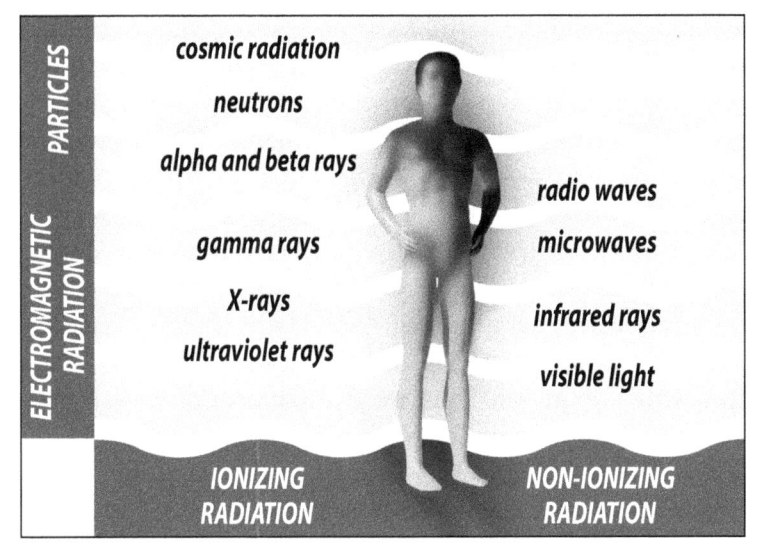

PARTICLES
- cosmic radiation
- neutrons
- alpha and beta rays

ELECTROMAGNETIC RADIATION
- gamma rays
- X-rays
- ultraviolet rays
- radio waves
- microwaves
- infrared rays
- visible light

IONIZING RADIATION | **NON-IONIZING RADIATION**

Figure 6. Different types of radiations to which man is daily exposed.

seashore or in mountains, on a sedimentary soil or a granitic one, near a coal or a uranium mine, at the Equator or the North Pole.

In particular, as this value is an average figure, one can already try to compare it to some extreme natural radioactivity values, as found in certain regions, to which local inhabitants are subjected daily. In some areas soil radioactivity can be close to zero while in others it can reach 1.5 mSv. In granitic areas, as well as in mountains, the annual average is around 3.0 to 3.5 mSv. In Brazil or in Iran, inhabitants of zones with naturally high concentrations of uranium or thorium can accumulate up to 100 mSv per year. Similarly, high concentrations of radon in closed areas such as not well-ventilated cellars can make doses climb up to 500 mSv per year. In terms of cosmic origin radiation, more precise measures have been made for people living in altitude. Mexico's 20 million inhabitants, located at 2,240 meters altitude, annually absorb 0.82 mSv, while this figure reaches 1.7 mSv for the 400,000 people living in Lhassa, Tibet (3,600 meters), and 2.0 mSv for the million Bolivians in La Paz (3,900 meters).

Let us remind ourselves at this stage that the difference between natural and artificial radionuclides only lies in the former's existence since the creation of the Earth, *i.e.* from "natural" origin, and in the latter's first man-made production with appropriate tools, *i.e.* since the end of the 30's. On the other hand, there is no difference between radiations of natural and of artificial origin, as they both have the same effect on inert or living matter, with results which do not allow any origin identification.

The discovery of new radioactive (artificial) isotopes is described in details in the history chapter. Radioactivity is a source of energy that, like all other energy sources (water, fire, charcoal, electricity, etc.) can be used by man for practical purposes or on the contrary, for its destruction power. Thus, the development of nuclear physics has lead to the creation of new weapons, as well as in parallel to the discovery of new and efficient treatments for patients affected with specific incurable diseases.

Today we have to take into account the fact that artificial radioactivity does exist, and therefore must be integrated in the calculation of the doses accumulated by man on a yearly basis. Thus, aerial nuclear weapon assays that were performed in the past now account for a yearly average of 0.10 mSv. The total of today's nuclear industrial activity, production of electricity included, add another 0.02 mSv in average. Nuclear medicine and medical imaging in general, including chest radiography, contribute for each individual to a dose of approx. 1.00 mSv. As a consequence, one has to add in average 1.10 mSv of artificial radioactivity to the above-calculated dose of natural radioactivity. As for natural radioactivity, this figure has to be modulated depending on whether one lives near a nuclear power plant or if one works in a radiology imaging unit.

In fact, when people aren't normally in direct contact with radioactivity as produced by the nuclear industry, the highest dose absorbed usually originates from medical examinations. A chest X-ray imaging gives an equivalent dose of 0.2 to 0.4 mSv to the patient. Technology improvements over the past forty years have reduced this dose by a factor of 20. In comparison, the dose inflicted by an abdominal scanner is about ten times higher while a dental radiography corresponds to approx. 0.002 mSv efficient dose per image.

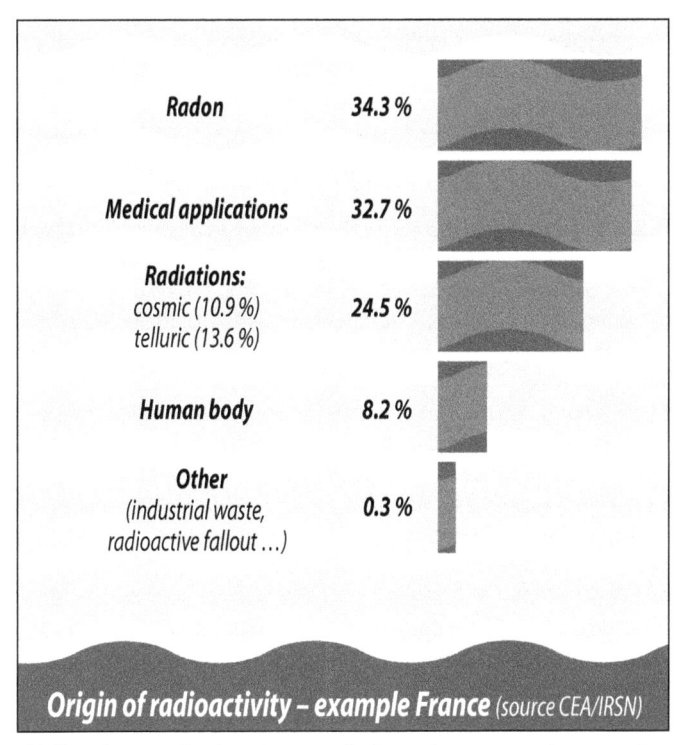

Radon	34.3 %
Medical applications	32.7 %
Radiations: cosmic (10.9 %) telluric (13.6 %)	24.5 %
Human body	8.2 %
Other (industrial waste, radioactive fallout ...)	0.3 %

Origin of radioactivity – example France (source CEA/IRSN)

Figure 7. Distribution of radioactivity to which man is permanently exposed.

As nuclear medicine is a technique which implies the injection of a radioactive substance, one may expect greater figures. However, those figures remain low because very small amounts of drug are sufficient to obtain the required image. Evaluating the levels of irradiation is a difficult task as it depends on the distribution of the drug throughout the different tissues and organs. The equivalent dose is estimated at a few millisieverts and remains below a dozen of millisieverts for a myocardial perfusion scintigraphy. In these cases as well, imaging technology improvements, in particular concerning detectors sensitivity, have since the 50's enabled a drug dosage as administered to a patient for a given examination to be divided by a factor of ten to one hundred.

On the other hand, the amount of doses that can be used therapy is limitless as it aims to destroy cancerous cells. Well-controlled methods of radiotherapy allow for the equivalent of tens of sieverts to be administered, thus confirming that high doses of radioactivity can save patient lives.

The Risks linked to Radioactivity

As with any source of energy, radioactivity presents risks that are obviously a function of the dose absorbed by an individual. Injecting a radioactive substance into a patient is not a trivial procedure. It is possible because tens of years of experience in this field have helped define the limit separating the beneficial effects from the harmful effects, even if this frontier still remains extremely imprecise.

The effects being a function of the dose, it is important to distinguish acute irradiation (accidental, radiotherapy, etc.) from chronic exposure (natural radioactivity, radioactive ambient environment, nuclear industry, etc.)

At this stage it is essential to remember that there is no difference between radioactivity that is natural in origin and radioactivity that is artificial in origin. This radiation, and therefore its effects, are one and the same. It is the quantity, the period of time, the energy, therefore the actual absorbed dose that makes the difference. By analogy, everyone agrees to say that UV radiation emanating from sunlight has the same benefits or damaging effects as that of artificial light.

At low doses, some researchers hypothesise that radioactivity would even be of benefit to a healthy person, but for the moment the scientific community has not yet succeeded in reaching agreement. The theory would have it that very weak levels of radioactivity contributed to the evolution process of animal species, and therefore man, and still participate today in the mechanisms of cell selection and the creation of self-defence systems via immune system stimulation. It must be acknowledged that scientific proof is very scarce due to the difficulties of being able to measure these weak doses and their impact. Only statistical studies covering a very large population can give any hints.

What is certain is that radioactivity causes mutations that lead to the formation of cancerous cells. The minimum dose required – the threshold – to generate a micro-tumour remains unknown. It is clear that it is not the initial radiation that is responsible for this evolution, because we are bathed in radioactivity on a daily basis. It should be noted that the average level of ambient radioactivity on our planet has decreased by a half since man has existed.

A parallel can again be made with this other form of radiation, sunlight: it is obvious that man needs the sun in order to survive (for example for vitamin synthesis via the skin), but over time, it has been proven that this same sun is also responsible for the appearance of melanomas. It is completely impossible to evaluate the minimum period of exposure to the sun required to generate a melanoma, and from that, the period of time that should not be exceeded in order to be sure of never being afflicted by this form of cancer. The answer depends on each individual and cannot be determined in advance.

Similarly, the effects due to radioactivity vary with each individual. And the actual observed effects are based on the minimum values (and not the average values) from which a disturbance has been noted. It is considered that for an adult, below 100 mSv over the whole body constitutes a weak dose. For an even exposure over the body, initial clinical effects have been noted from 250 mSv in the form of nausea and a slight drop in the number of red blood cells. Blood count is changed above 1,000 mSv and one person in two will die if the dose exceeds 2,500 mSv. In the case of local irradiation, the effects depend on the organ or tissue affected and the cells in direct contact (skin and eyes in the case of irradiation, skin in the case of contamination, lungs in the case of inhalation) will be the most affected.

These values serve as a reference for physicians treating patients. In the case of imaging, doses remain well below the theoretical risk values. In the case of therapy, the rules are different, since the aim is to destroy the supernumerary or cancerous cells. Very strong doses may be applied as they alone can lead to healing. The selectivity of the vector, the reduced retention time in healthy cells and the speed of elimination are all factors that contribute to the maximum reduction of the risk of causing a new cancer.

For staff working in a radioactive environment it is the ALARA rule (As Low As Reasonably Achievable) that takes precedence, and which is applied in all sectors of the nuclear industry, in the radiopharmaceutical industry or in the nuclear medicine departments of hospitals. For, in the end, the most exposed people are not those who undergo an injection of a radioactive substance once or twice in their lives, but actually those who are in daily contact with these substances or with patients undergoing radioactive treatment.

III. Radionuclides for Nuclear Medicine

Isotopes used in nuclear medicine are characterised by their short half-life that must take into account imaging constraints (acquisition time) and the biological elimination process (effective presence in the human body), as well as production, logistics and waste treatment constraints. When a half-life is shorter than two hours, it is not seriously conceivable to run a production and a distribution phase of the drug without huge product losses. Half-lives which are higher than four days raise new problems in terms of storage and waste treatment. These drugs must be injected, swallowed or inhaled, and the elimination rate from the body or the decay must be sufficiently fast to avoid any negative consequences on healthy human tissue. Finally, their energy must be adapted to the needs of both the patient and the physician.

Gamma Emitters (γ)

The following table regroups the most commonly used gamma-emitting radionuclides in nuclear medicine. The values in brackets in the column "Major energies" correspond to the total energy of disintegration. In practice, Technetium 99m, Iodine 123 and Thallium 201 are today the most frequently used radionuclides in nuclear medicine imaging modalities.

Radionuclide	Half-life	Types of rays	Major energies	Comments
Chromium 51	27.7 d	X, γ	0.32 MeV	
Gallium 67	3.26 d	EC γ	93 keV 0.19 MeV	Imaging, but potential therapeutic use due to its Auger electron emission
Indium 111	67.3 h	γ, EC	0.17 MeV 0.25 MeV	Nuclide that can potentially substitute Yttrium for imaging due to its similar chemistry. Potential therapeutic use due to its Auger electron emission (oncology – haematology)
Iodine 123	13.2 h	EC, γ	0.16 MeV	Ideal radionuclide for covalent labelling of organic molecules; potential therapeutic use due to its Auger electron emission (oncology – neurology)
Technetium 99m	6.02 h	γ	0.14 MeV	Most common radionuclide for diagnosis, produced on site from the Molybdenum 99/Technetium 99m generator (oncology – cardiology)
Thallium 201	3.05 d	X, γ	0.17 MeV	Cardiology imaging
Xenon 127	36.4 d	X, γ	0.20 MeV	Radioactive gas for imaging (oncology – pneumology)
Xenon 133	5.24 d	β^-, γ	81 keV	Radioactive gas for imaging (oncology – pneumology)

EC: Electronic Capture

Positron Emitters (β⁺)

The most frequently used positron emitters radionuclides in nuclear medicine are given in the following table. Among these radionuclides, Fluorine 18 is the only one currently available on an industrial level. All others are mainly research tools. The disintegration of β⁺ emitters always goes with the production of two 511 keV gamma rays, corresponding to positron annihilation (not in the table). Energy values as provided between brackets correspond to the total energy of disintegration.

Radionuclide	Half-life	Types of rays	Maximal energies	Comments
Carbon 11	20.4 min	β⁺	0.96 MeV	Half-life just sufficient to label an organic molecule and to inject to man immediately following labelling
Fluorine 18	110 min	β⁺	0.63 MeV	Ideal radionuclide for PET imaging; requires however proximity with dedicated cameras (oncology – haematology – neurology)
Gallium 68	1.12 h	β⁺ γ	1.90 MeV	Positron emitter that can be produced from a Germanium 68/Gallium 68 generator
Iodine 124	4.18 d	β⁺, EC γ	2.13 MeV	Positron emitting radionuclide with a strong γ emission
Nitrogen 13	10 min	β⁺	1.20 MeV	Positron emitter used only in research mainly under the form of ammonium ion
Oxygen 15	2 min	β⁺	1.73 MeV	Radionuclide with a too short half-life forbidding integration in a molecule; therefore used in the form of labelled water directly from the cyclotron
Rubidium 82m	6.4 h	β⁺ γ	3.35 MeV	Radionuclide obtained from a Strontium/Rubidium 82m generator; used in cardiology
Technetium 94m	53 min	β⁺ γ	2.47 MeV	Positron emitter radionuclide currently used as research tool
Yttrium 86	14.7 h	β⁺	1.25 MeV	Positron emitter radionuclide allowing imaging of equivalent Yttrium 90 labelled therapeutic molecules; currently research tool only

3 Electron Emitters (β⁻)

The number of β⁻ emitting radionuclides is particularly important. At this stage, about a dozen elements present more interesting physical characteristics. The most often used radionuclides are listed in the following table. The interest of some of them is already declining (Phosphorus 32 for example), while new ones are emerging (Holmium 166 and Lutetium 177). In practice, Samarium 153, Strontium 89 and Yttrium 90 are the most often used ones.

Radionuclide	Half-life	Types of rays	Maximal energies	Comments
Copper 67	61.9 h	β⁻, γ	0.39 MeV	
Erbium 169	9.4 d	β⁻	0.35 MeV	Radionuclide for radiosynoviorthesis (rheumatology)
Holmium 166	26.8 h	β⁻	1.85 MeV	Therapy (oncology)
Iodine 131	8.02 d	β⁻, EC, γ	0.61 MeV	Diagnosis and therapy, ideal for the treatment of thyroid diseases (oncology)
Lutetium 177	6.71 d	β⁻, γ	0.50 MeV	(oncology)
Phosphorus 32	14.3 d	β⁻	1.71 MeV	(haematology)
Rhenium 186	3.77 d	β⁻, X, γ	1.08 MeV	Radionuclide for radiosynoviorthesis (rheumatology)
Rhenium 188	16.9 h	β⁻, γ	2.12 MeV	Radionuclide for therapy produced from a Wolfram 188/Rhenium 188 generator (oncology)
Samarium 153	46.3 h	β⁻, γ	0.70 MeV	Used mainly in the palliative treatment of pain linked to bone metastases (oncology)
Strontium 89	50.5 d	β⁻	1.49 MeV	Used mainly in the palliative treatment of pain linked to bone metastases (oncology)
Yttrium 90	64.1 h	β⁻	2.28 MeV	Radionuclide produced either by irradiation in a reactor, or from a Strontium 90/Yttrium 90 generator. Radionuclide for metabolic radiotherapy (oncology – haematology) radiosynoviorthesis

4 Alpha Emitters (α)

Several teams of researchers are currently working on the development of molecules labelled with alpha emitters. In the series of the actinides, most of the radionuclides are alpha emitters, uranium and plutonium being the best known ones. Unfortunately, only very few of these radionuclides do have properties that fit with nuclear medicine constraints. The legislation currently in vigour also limits their use and particularly their transport. So far only one product is commercially available: Radium 224 is used in the therapy of ankylosing polyarthritis. The following table regroups the alpha emitting radionuclides that may be marketed within the next tens of years.

Radionuclide	Half-life	Types of rays	Maximal energies	Comments
Actinium 225	10.0 d	α	5.73 MeV	Alpha-therapy; produces 4 α particles per atom during the process of disintegration. Parent isotope used in the Bismuth 213 generator (oncology – haematology)
Astatine 211	7.2 h	α	5.87 MeV	Alpha-therapy
Bismuth 212	60.6 min	α, β⁻	6.05 MeV	Alpha-therapy
Bismuth 213	45.6 min	α, β⁻	5.87 MeV	Alpha-therapy, obtained in a generator based on Actinium 225 decay
Radium 224	3.64 d	α	5.68 MeV	Precursor of Bismuth 212 in a generator (rheumatology)

5 Radionuclides for Brachytherapy and External Radiotherapy

The following radionuclides are used in external radiotherapy, internal radiotherapy (temporary implants) or in brachytherapy (permanent implants). They are definitely out of the scope of nuclear medicine. They are usually characterised by long half-lives and used as a source of radiation. In consequence, such radionuclides must not be injected.

Radionuclide	Half-life	Types of rays	Maximal energies	Comments
Californium 252	2.6 years	n	6.11 MeV	Source of neutrons for the techniques of therapy based on neutron capture
Caesium 131	9.7 d	γ, CE	34 keV	Source of irradiation – sterilisation
Caesium 137	30 years	β⁻, γ	1.17 MeV	Source of radiotherapy
Cobalt 60	5.26 years	β⁻ γ	0.32 MeV 1.17 MeV 1.33 MeV	Source of radiotherapy (cobalto-therapy)
Iodine 125	59.9 d	γ, CE, X	27 keV	Isotope used in biomedical analyses (radio-immunoanalysis) of blood and urine. Used also in permanent implants for prostate cancer therapy
Iridium 192	73.83 d	β⁻, γ, X	0.67 MeV 0.32 MeV	Implants used in the form of metallic wires – brachytherapy – breast cancer
Palladium 103	16.99 d	X, γ	20 keV	Radioactive implants for the treatment of prostate cancer

6 Other Radionuclides

This last table lists various isotopes that are of some interest to the industry, and sometimes also to medicine, but that have nothing whatsoever to do with radiopharmaceuticals. Some of them are filiation isotopes; it is interesting to know that they exist, as they can be part of nuclear waste from nuclear medicine isotopes. All of them are listed for information purposes, and their properties can be compared to those of the other radionuclides (energy and half-life).

Radionuclide	Half-life	Types of rays	Major energies	Comments
Carbon 14	5,730 years	β⁻	0.16 MeV	Used in object dating; often used to study animal metabolism, and more rarely, human metabolism; present in all carbon containing material, and hence in human food
Hydrogen 3 (Tritium)	12.3 years	β⁻	19 keV	Present in nature in trace amounts, therefore also in the human body. Mainly used in research as a biological tracer

SOME BASIC NOTIONS OF RADIATION 51

Radionuclide	Half-life	Types of rays	Major energies	Comments
Molybdenum 99	2.75 d	β⁻ γ	1.37 MeV 0.74 MeV	Radionuclide that cannot be used directly as a tracer, but very important as it is the source of Technetium 99m in the Molybdenum 99/ Technetium 99m generator
Strontium 90	28.5 years	β⁻	0.55 MeV	Precursor of Yttrium 90 in the Strontium 90 / Yttrium 90 generator
Technetium 99	214,000 years	β⁻	0.29 MeV	Decay product of Technetium 99m. Technetium does not exist in the form of a stable isotope
Tungsten 188	69.4 d	γ, β⁻	0.35 MeV	Precursor of Rhenium 188
Uranium 235	710 millions years	α	4.40 MeV	
Uranium 238	4.5 billions years	α	4.20 MeV	

Summary

In this chapter we have learned that a **radionuclide** is a substance that degrades with time in a very constant manner, and emits one or several **radiations**. This degradation or **decay** is defined by a constant, the **period** (or **half-life**) corresponding to the time it takes for half of the remaining substance to disappear. This half-life is specific for each radionuclide.

The type of emitted radiation is also specific for each radionuclide. There are four types of radiation which are of interest to nuclear medicine: for diagnosis purposes, **gamma rays (γ)** and **beta plus radiations (β⁺)** have respectively led to the development of the imaging technology of **SPECT** (Single Photon Emission Computed Tomography) and **PET** (Positron Emission Tomography).

Beta minus (β⁻) or **alpha (α)** radiations are used in metabolic radiotherapy.

The amount of radioactivity emitted by a radionuclide is expressed in **becquerels (Bq)**, one Becquerel corresponding to the degradation of one atom per second. The amount of energy transferred by this radiation is expressed in **grays** (**Gy**) while the equivalent absorbed dose by a tissue is given in **sieverts** (**Sv**).

Most of the matter existing on Earth, including living matter and hence the human body, contain radioactive substances.

The total amount of natural radioactivity absorbed by one man over the course of a single year corresponds in average to 2.4 mSv. This value becomes a reference when it is compared to the radioactive doses received from non-natural sources.

Among all the radionuclides with potential in nuclear medicine, we shall remember the currently most used ones: Iodine 123 and Technetium 99m as γ emitters, Fluorine 18 as β^+ emitters, Iodine 131, Strontium 89, Yttrium 90 and Samarium 153 as β^- emitters.

CHAPTER IV

Gamma Ray Imaging

Nuclear medicine is above all a functional imaging tool. Monitoring the distribution of radioactive substances injected into patient gives physicians information that is not accessible by other methods in fields as diverse as oncology, haematology, cardiology and rheumatology. In fact, almost all medical specialties can benefit from this technology. In addition, it is clearly evident that the ionising properties of radionuclides can also be of unique benefit to the patient during therapy. This aspect will be dealt with in one of the following chapters. For almost half a century quality images could only be obtained with gamma radiation and most radiopharmaceutical products were developed on the basis of gamma emitting radionuclides associated with the imaging tools available for this type of radiation. Positron sources have been developed in the course of the last two decades only following the invention of dedicated cameras and above all more user-friendly methods of producing β^+ radionuclide sources. We'll take a closer look at this fast-developing technology further in this book.

Methods of Imaging

High quality anatomical or morphological (the shape of an organ) images may be obtained via three different methods, besides scintigraphy: radiology (XR), ultrasound (US) and magnetic resonance imaging (MRI). The four techniques are based on completely different principles and, as a result, various and often complementary information can be obtained. These imaging techniques are said to be non-invasive and do not require either anaesthesia or hospitalisation. They provide much more than a simple three-dimensional photograph of tissue and organs which are not visible from the outside.

Radiology and X-rays

The use of X-rays for obtaining medical images is the oldest method known. An X-ray is an electromagnetic wave of the same type as light waves, but with a higher energy level, thus capable to some extent of travelling through matter. The technique is based on matter's capacity to attenuate an external beam of X-rays depending on whether it is solid, liquid or gaseous. Therefore, organs will allow radiation to pass through them according to their density, thickness and constitution; the intensity of the rays is measured on photographic film or using a specific detector. It is possible to render certain areas opaque by injecting a contrast product into the cavities (digestive system for example), thus holding back the X-rays.

Initial applications focussed on organs and tissues presenting a different coefficient of absorption. Reference should be made to the most common application, displaying the skeleton, and allowing all the defects and malformations to be seen. Image quality improved with the quality of equipment and detection methods. Today's images are nothing like those obtained in the late 50's. Not only have the resolution and quality considerably improved, but above all, the quantity of radiation necessary to obtain an image is much lower nowadays, therefore reducing the doses absorbed by patients, and also by the operator. A gain of a factor of 20 has been achieved in the doses received during a simple lung X-ray during the last 40 years.

Even if most of the images that are still produced are a silver grey, flat and rigid, the technology was revolutionized by the advent of digital imaging. Thanks to the increase in computers' calculation power, the time needed to process images has been reduced considerably.

Scanning, another name for tomography (from the Greek *tomein* meaning "to slice"), has improved the quality of results. Computed tomography uses an X-ray source which turns around the body to give a sectional image. A helicoidal or spiral scanner combines this rotation with an axial movement, allowing a complete three-dimensional image of the body to be taken in 20 seconds with the latest generation devices. In the case of a three-dimensional scan of the lungs, patients only have to hold their breath for 7 to 8 seconds. The matrix detector that equips these devices enables tumoral and inflammatory processes to be monitored in all the body's organs. Once the data has been acquired, the software equipping these tools gives the radiologist the option of navigating from one organ to another, and of isolating a particular element in order to be able to focus on the element of interest. Then, the entire data, including the images, can be sent to a colleague for confirmation or additional expertise with a simple click of the mouse.

At the same time, contrast agents used in radiology have made considerable progress, adding to the quality and resolution of the images. As soon as it was understood that the density of the matter crossed by the ray was linked to the opacity of the image, it was also understood that a specific image could be taken of irrigated regions, as long as a contrast liquid could be injected into them. A contrast medium that was neutral, therefore non-toxic, and which could be eliminated rapidly still needed to be found. Blood circulation slides could be taken with initial applications, followed by liver or bile function images. Contrast agents, often based on iodine or barium, have now become unavoidable, as they give an extremely precise display of certain anatomic details such as very small vessels for example.

Radiology is more particularly used in orthopaedics, rheumatology and orthodontics. It is also very informative in pneumology (preventive radiology of the lungs or where there is a suspicion of lung cancer) and in oncology (mammography, preventive examination for the detection of breast cancer). Contrast radiography is more specifically used in gastroenterology (stomach, liver, gallbladder, intestines). Scanners are ideal to obtain three-dimensional images (cardiovascular system, detection of tumours, analysis of organs before surgery, etc.)

A patient that has just been subject to a radiological examination is not radioactive, even if X-rays are a form of radioactivity. Outside the field, the effect of the radiation disappears. Progress in these last years has effectively aimed at reducing both the time taken to acquire the image and the dose necessary for this acquisition, while still maintaining, even improving, quality. But whatever the technology supporting X-ray detection may become, this method will remain a diagnosis procedure and will presumably be limited to anatomical imaging.

Ultrasounds

Ultrasounds are waves that are imperceptible to the human ear, but which retain such properties as reverberation (echo) or matter absorption (attenuation). By taking advantage of these two characteristics and the properties of tissue subject to ultrasound, it has been possible to develop a tool capable of measuring and analysing the nature of its reflection according to the tissue through which it has travelled and off which it has bounced. Measurement of the time required for the wave to be detected also permits the distance travelled to be calculated. The ultrasound technique permits restitution, thanks to appropriate software, of the shape of organs reached by the beam.

The technique has become well-known in monitoring pregnancy, but it was also developed in other fields ranging from cardiology to oncology. Doppler ultrasound enables fluid flow (blood in particular) and tissue irrigation to be analysed.

Image definition has been very clearly improved, and colour, as well as animated sequences, have made their appearance. Several contrast agents, composed of air micro-bubbles trapped in biodegradable substances, have been developed. As ultrasounds are completely reflected by air, these products allow images to be taken of cavities in the same way that contrast agents are used in radiology. The resolution of traditional devices seems low, but the information obtained is sufficient for most analysis, avoiding investment in a more powerful device. The latest, top-of-the-range tools offer a resolution lower than a tenth of a millimetre and are used more particularly in eye and skin analysis.

Due to the opacity of areas containing air, the lungs, digestive tract and the bones cannot be analysed using ultrasounds. This technique is particularly useful in cardiology (heart and vessels, despite difficulties of interpretation due to interference from the ribs), gastroenterology (liver, gallbladder, kidneys), urology (genitals, bladder) and of course, in obstetrics.

A patient who has just been subject to an ultrasound examination is, of course, not radioactive. But as with X-rays, ultrasounds can only be used as a diagnosis procedure limited to anatomical imaging. Some teams of researchers are looking for contrast agents attached to vectors targeting a specific mechanism, but are far from finding an agent that can be marketed.

Magnetic Resonance Imaging

In creating the name Magnetic Resonance Imaging (MRI) physicians wanted to avoid using a word that might frighten patients, the true name of the technology being Nuclear Magnetic Resonance. Nuclear physicians, who have no connection with MRI, have not raised this type of question. On the other hand, no consideration of radioactivity comes into the MRI concept, and it only refers to magnetism and magnetic field measurement. Subject to a magnetic field of a defined intensity, the nuclei of certain atoms, themselves being able to be considered as mini-magnets, align themselves to this field. By subjecting these atoms to a wave of very short radio frequency and specific to this atom, one of their parameters can be changed, their direction of rotation or spin. They enter into resonance with this wave, that is to say, they start to vibrate at the same frequency. When this external radiofrequency is switched off, their return to a normal state is expressed by the restitution of this energy in the form of a signal which can be captured and measured. The origin of the signal can be found practically atom by atom as these are located in space. In organic tissue, hydrogen is the most common atom with this property. Hydrogen is also the majority atom in human tissue because it is the main element in water and fats, matter making up almost nine tenths of our body weight. Therefore, three-dimensional displays of the body density of water, fats and other organic matter containing hydrogen can be obtained with MRI. As this density differs from one organ to another, it is easy to obtain the outline of these elements with precision to within a few millimetres. Much more powerful tools are being developed that will give a resolution of under a millimetre. As the technique is still costly, it is preferable to use traditional radiological methods to display a broken arm, and the result is just as convincing.

Hydrogen is by far not the only element that could be used in MRI, but the density of its atoms remains the greatest. Iron is an interesting element on the other hand. It is well known to physicians, above all because the presence of particles or pieces of metal (implants, prostheses), even powders and iron oxides (contained in makeup products) can lead to major distortion of the final image, impeding its interpretation.

Contrast agents developed in this field are based on the properties of gadolinium, another metal with interesting magnetic properties, which gives specific organ imaging when grafted onto certain substances.

Functional analysis is possible in some very specific cases, like brain irrigation or heart functioning monitoring, via live analysis of the rate of oxygen contained in the blood.

If MRI was not as expensive and was more widely available, it would easily replace most of the techniques described above. For the time being, it is kept for examinations for which diagnosis seems more difficult (soft tissue such as muscles, tendons, brain, tumours) or in indications for which this method is much more effective or unique (neurology, ophthalmology, cardiovascular, endocrinology, oncology, etc.)

A patient that has just undergone an MRI examination is no more radioactive than a patient that has received ultrasounds, nor are they magnetised or magnetic. But as with X-rays and ultrasound, for the moment MRI can only be used as a diagnosis procedure limited to anatomical imaging. In this field as well, it is hoped that contrast agents will be found for functional imaging use in oncology, but the sensitivity of a radiopharmaceutical remains technically inaccessible.

A Combination of Techniques

The three procedures described above are, and will remain for a long time to come, limited to anatomical analyses of results. An organ can very easily be identified, its conformation checked, changes in its shape can possibly be monitored over time, and particularly its growth or reconstitution, but physicians are not capable of finding out if the tissue being studied is functioning optimally. No information on the mechanisms controlling the cells is available. Only a method which could imitate these cell mechanisms could give an answer. Nuclear medicine imaging is the only method capable of carrying out both a morphological analysis of the organs and a functional analysis of tissue. Its precision of measurement remains an advantage over the first three methods. Physicians have understood this, and they also know that the four methods are complementary, each providing information at a different level.

I. Nuclear Medicine Imaging Methods

Nuclear imaging uses either salts and radioactive complexes or traditional pharmaceutical molecules (drugs) or biochemical molecules (hormones, antibodies) that we will call vectors, to which a radionuclide (label) is grafted. These molecules are introduced into the organism by injection, ingestion or inhalation and the physician can follow their distribution with appropriate detectors (cameras) thanks to the radioactive signal emitted by the radioisotope (scintigraphy). The words tracer or radiotracer are used to define these radiopharmaceutical molecules, which are used for obtaining images.

The whole structure of the molecule must be designed in such a way to ensure that the radioisotope does not break free from its vector through a natural biological process (rapid metabolism), or that it is not eliminated too quickly from the organism. This entity participates in the natural cell mechanism and allows for it to be monitored. For nuclear medicine therefore, diagnosis is above all based on functional exploration. The distribution of the product and its associated image change over the course of time, until the tracer is fully eliminated. This technique cannot be used to obtain images of an organ that has been damaged, since the radioactive molecule must be able to participate in the cells' biological mechanisms. It could for example be used to distinguish a brain that is alive from one that is clinically dead. This differentiation is not possible with other imaging techniques. MRI, X-rays and ultrasound will still provide an image of the state or form of a body or an organ, whether the individual is alive or dead. However, images of an organ that is no longer functioning will be totally different if they're obtained with a radioisotopic tracer, even if the said organ is irrigated artificially.

The quantities administered are extremely low, but sufficient to be detected. The great strength of imaging in nuclear medicine lies in this sensitivity. Gamma cameras are able to detect minute quantities of tracer, less than a billionth of a gram, while X-rays require a local concentration the order of a tenth of a milligram, and MRI as much as a tenth of a gram, to produce a contrast. The quantities of tracers required for diagnosis studies in nuclear medicine are so low that although they participate in the normal biological mechanism of a cell, they do not interfer with it. The quantity of radioactive iodine (Iodine 123 or Iodine 131) used during thyroid scintigraphy remains more than a thousand times smaller than the dose of stable iodine (Iodine 127) absorbed with the daily food intake.

By contrast, this high degree of sensitivity is accompanied by poor spatial resolution, that is to say it is difficult, quasi-impossible even, to obtain clear images and visible details. The shape and size of small lesions cannot be observed precisely, and as of now the method does not allow two neighbouring lesions of less than a centimetre in size to be distinguished. However, even very small tumours will remain visible if the contrast is sufficient, in other words if there is not too much background noise, or better still, if the label is truly specific to the tumour concerned and does not fix in surrounding tissue. This physical limit explains why the technique is not used routinely for certain applications, such as in scintimammography. From the point of view of physical properties, *i.e.* at the detectors' level, the tool has doubtless reached its limits with resolutions of approx. 3 to 5 mm. The recent arrival of the mixed PET/CT technique now enables physicians to break through this limit, since the location of the source has become extremely precise, even if the image of the source remains less than clear. Another great step forward needs to be made, by improving the specificity of the vectors, that is to say, their capacity to target a tumour while remaining at a very low concentration level in healthy cells.

The radioisotopes used in medical nuclear imaging need to have a relatively short half-life. As a result, the patient's own radioactivity level decreases rapidly over time. When this level of radioactivity has lowered to within public safety standards, patients can leave the hospital. In fact, in most diagnosis uses, this safety threshold is never even exceeded, meaning that no hospitalisation or confinement to a protected hospital department is required. It should be remembered that the level of radioactivity is reduced by a factor of about 1,000 after a period of 10 half-lives and that radioactive emissions are 100 times lower at 1 m than at 10 cm. Therefore, spending less time with, and staying further away from, radioactive patients considerably reduce exposure risks. This rule is especially valid for medical staff working closely with patients. The doses emitted following a simple injection of radioactive imaging substances are of no consequence for the patient, and therefore are of even less consequence for the patient's friends and family.

Scintigraphy

Scintigraphy covers techniques analysing γ rays that are transformed into a flat image, similar to an X-ray plate. The radiation comes from the substance injected into the patient and which has concentrated in certain tissues. Therefore, these tissues become themselves the source of this radiation, which is recorded on the photographic plate or specific detectors, in contrast to X-ray imaging where an external radiation source is required during the period of image acquisition.

Let us take a quick look at the diseases that can benefit from scintigraphy. Almost all organs can be investigated by this technique, and above all their function can be checked without the need for a surgical operation or biopsy. This method is said to be non-invasive.

One of the most frequently performed investigations is probably bone scintigraphy. A whole body image of a patient suffering from bone pain provides information on bone function. Scintigraphy will detect any increase in bone metabolism corresponding to a lesion. Almost all solid cancers eventually form metastases which systematically develop on the bones. Bone scintigraphy enables practitioners to search for these bone metastases in an effective manner.

The heart is another organ that benefits from advances in this technology. After any heart discomfort or in almost all cases of chest pain where there is, therefore, a suspicion of heart failure, the patient is subject to a test that frequently requires myocardial perfusion scintigraphy. This tool allows testing of the heart muscle's condition as well as the way in which it is being irrigated. In other words, it allows testing of the heart pump's functioning. The technique is based on monitoring the heart's biological mechanisms and does not dwell on the shape of the organ. In consequence myocardial perfusion scintigraphy images show very little resemblance to a heart.

The kidneys also benefit from the properties of specific radiolabelled vectors. Renal scintigraphy provides the urologist with important information concerning blood exchanges and the organ's functionality. Patients suffering from hypertension and diabetes, as well as patients likely to suffer from kidney stones are subject to routine tests.

Scintigraphic imaging of the thyroid is the most widely known specific method due to the systematic accumulation in this tissue of any iodine absorbed. Thyroid scintigraphy is used to screen for abnormal functioning of the thyroid gland (hyper- or hypothyroid), as well as malignant tumours of this gland. Secondary thyroid cancers accumulate iodine in the same specific manner, thus it is possible to monitor the disease's development using the same tools.

As for the brain, the tools available to date allow the condition of irrigation vessels to be tested, and diseases such as epilepsy, and possibly Alzheimer's to be monitored. More specific molecules for brain scintigraphy, *e.g.* for early detection of Parkinson's disease or Alzheimer's are still at an early stage of development. Nevertheless the products that were first used in neurology, and more particularly for monitoring neurodegenerative diseases, will soon be available on the market.

Pulmonary scintigraphy is a more particular technique. It requires the use of a radioactive gas that is inhaled, giving an overall image of all the alveoli in the accessible airways. This technique is called pulmonary ventilation scintigraphy and supplements pulmonary perfusion scintigraphy, in which a lung image is obtained on the blood vessel side after injection in the veins. These two images, which complement each other, show the interface between the oxygen in the air and the blood system in the lungs. For a patient suffering from a pulmonary embolism, an incomplete image pinpoints the areas no longer participating in the oxygenation.

Finally, nuclear medicine can provide vital indications in determining infected or enflamed areas, when these are touching soft tissue or internal organs that are difficult to access. For example this technique, very little used in Europe but increasingly in the United States, can confirm appendicitis before an operation. This diagnosis aid helps physicians avoid post-operative legal complications, if it turns out that an operation should not have taken place.

2 The Products used in Scintigraphy

The diagnosis radiopharmaceuticals, which concentrate in the tissue to be analysed, possess particular biological, chemical and radiological properties. Experience gained over the course of the last fifty years gives a better understanding of the mechanisms that favour this local concentration, thus enabling researchers to faster develop more specific, and therefore more effective molecules.

The diagnosis effectiveness of a radiopharmaceutical is obviously linked to the specific characteristics of the vector, the non-radioactive part of the molecule and to the half-life of the isotope that is attached to it. These parameters are not sufficient however, because if this radiopharmaceutical molecule degrades before it reaches its target or leaves it with no interaction, it becomes of no use. Another

Commercially available diagnosis radiopharmaceuticals and their indications (gamma emitters)

Isotope	Chemical form of the radiopharmaceutical	Indications
Chromium 51	Sodium chromate	In vitro/ex vivo labelling of red blood cells (measure of volumes, masses and survival time)
	Edetate	Renal filtration
Gallium 67	Citrate	Tumoral imaging, inflammation localisation
Indium 111	Chloride DTPA Oxyquinoline	Labelling of peptides, proteins and antibodies for oncology and haematology imaging
Iodine 123	Sodium iodide	Thyroid scintigraphy Labelling of molecules for imaging
	Iobenguane (MIBG)	Morphological and functional imaging of the thyroid Detection of tumours
	FP-CIT	Neurology (Parkinson disease)
	Fatty acid	Cardiac metabolism studies
Iodine 125	Iodinated human albumin	Blood volume and blood albumin renewal studies
Iodine 131	Sodium iodohippurate	Renal filtration studies
	Iodomethylnorcholesterol	Adenocortical diseases
Iron 59	Iron citrate	Gastro-intestinal absorption
Rubidium 82		Cardiac imaging
Technetium 99m	Human albumin	Vascular and pulmonary imaging
	Bicisate	Brain imaging
	Disofenin	Evaluation of the biliary function
	DTPA	Vascular cerebral, renal and pulmonary imaging
	Exametazine (HMPAO)	Cerebral perfusion – labelling of blood cells for the detection of infections
	Mebrofenin	Liver imaging
	Mertiatide	Renal filtration
	Sodium pertechnetate	Vascular, cerebral imaging; imaging of the salivary glands, the gastric tract, the lachrymal tract
	Phosphonates (medronate - oxydronate - pyrophosphate)	Bone scintigraphy – imaging of bone metastases
	Phytate	Liver imaging
	Tin pyrophosphate	Vascular imaging
	Sestamibi	Cardiac imaging
	Colloidal rhenium sulphide	Liver imaging, sentinel node detection
	Succimer	Renal cortex imaging
	Tetrofosmin Antibody and peptides	Infectious sites from tumoral structures imaging
Thallium 201	Chloride	Myocardiac scintigraphy (detection of infarcts, ischaemia and necroses)
Xenon 127 et 133	Xenon	Pulmonary and cerebral perfusion

Abbreviations: DTPA, diethylene triamino pentaacetic acid; HMPAO, hexamethyl propylene amine oxime; FP-CIT, fluoropropyl carbomethoxy iodophenyl tropane; MIBG, metaiodo benzyl guanidine.

parameter called the effective half-life needs to be considered, therefore, which takes into account the contact time between the ligand and its target. More important still than knowing the radioisotope's emission time, it is crucial to have data available concerning the time of the molecule's residence in the target cell, the biological half-life. The ideal would of course be to have access to a molecule with a maximum residence time in the cell or target organ, and almost none in all other parts of the body. This is never the case. With the exception of iodine used for imaging or treating the thyroid and for which the isotope itself is the ligand, it is actually the vector which plays this important role, which consists of selectively targeting the radiopharmaceutical onto the target organ. The combination of the half-life of the isotope and the biological half-life defines the effective half-life, which itself determines the real impact of a molecule on its target.

The ideal radiopharmaceutical is defined by a high level of binding (the capacity of a molecule to attach itself on the inside or on the surface of a cell, *i.e.* on a receptor). For a specific receptor-ligand pair, most of the time, these fixation values are known. The radiochemist's main task will be to attach the radioisotope to this molecule while avoiding any disturbance to its interaction with the receptor, and therefore to its biological effectiveness.

II. Imaging Tools

The images produced in nuclear medicine are essentially obtained using a camera that detects gamma rays. This specific gamma imaging tool is equipped with a detection head that analyses an area up to 40 x 60 cm in a single pass. The rays, which are emitted in every direction within the area, are selected by a collimator as they pass through, and only those which are perpendicular to the detector are taken into consideration. This detector composed of a crystal that is sensitive to radiation (sodium iodide for example) is coupled to a photomultiplier which transforms the impact of the radiation into an electronic impulse. The impacts are analysed point by point in a planar image. The quality of a gamma camera depends on the the detection crystal's sensitivity and the collimator's resolution. Moving the detection head along the length of the body gives a static planar scintigraphy image of the whole body within minutes. This technique is used in almost all types of nuclear medicine indications, with the exception of those concerning the heart and the brain.

The same type of equipment, used over a specific region, gives a dynamic image with which the evolution of the radiopharmaceutical's distribution through the organs can be monitored over a precise length of time. The processes of blood irrigation can also be monitored. Finally, observing the functioning of the liver or kidneys is no longer a problem when plates are taken sequentially, over a predetermined period and with defined and regular spacing between each acquisition.

The SPECT method (Single Photon Emission Computed Tomography) applies the gamma camera principles to a scanner, a tool that is equivalent to the devices used in radiography. The photon source is by contrast located in the patient himself since the radioisotope has been injected. This method does not, of course, require any additional irradiation. The equipment generally uses two or three detectors which revolve around the patient thus giving a cross-sectional image. If this acquisition of data is carried out in parallel with a linear analysis, a three-dimensional image of the body is obtained. This recent improvement has been made possible thanks to the development of very powerful computers with which these devices are equipped. The SPECT method is ideal for analysing areas that are well-defined and limited in size, such as the heart or brain.

As acquisition times remain relatively long, moving organs such as the heart are more difficult to image, or at least to interpret. As a solution to this problem, images are taken according to the heart rhythm. The period between two beats is divided into thirty sequences and the gamma impacts recorded during each fraction of time are accumulated separately. After ten minutes, thirty different images, each corresponding to a precise fraction of the heart beat, are recorded. By displaying them in order and in a loop, an animated sequence of the heart beat is reconstructed. This dynamic mode technique is called gated SPECT.

This special technique may also be used when images have to be taken in the neighbourhood of the lungs, which are also subject to regular movement. The acquisition of images can be synchronised with breathing movements.

The PET camera (Positron Emission Tomography) benefits from the advantage provided by the simultaneous emission in two opposite directions of two gamma photons, when the positron collides with an electron. The detectors are placed in a ring around the patient and measure the two concomitant double impacts. Their origin can be deduced with precision with mathematical analysis. The equipment is sufficiently developed to obtain three-dimensional images of the whole body, and the precision is such that this technique has become common in brain analysis. But the most common application is linked to the FDG vector which enables the detection and precise localisation of tumours and their metastasis. The next chapter looks at this most recent technology in more detail.

III. Detection of the Sentinel Node

Among the other tools used for detecting radioactivity, we will only mention peroperative probes. These permit ad-hoc localisation of the concentration of a radioactive substance which was previously injected into the patient, and which has the property of accumulating specifically in the target zone. The device, the size of a large pencil, enables the surgeon to find a zone of strong radioactivity concentration using this detector. In this type of surgery, radioactivity plays the role of a label equivalent to a dye, the detection of which can only be achieved with an appropriate tool.

Two types of probes are available: scintillation detectors using a scintillation crystal (sodium iodide activated with thallium) coupled to a photomultiplier and semi-conductor probes (cadmium-tellurium) linked to a preamplifier. These probes are used for detecting and measuring concentrations of radioactivity in tumours (colorectal cancers, breast cancers, ovarian cancer, melanomas) and for detecting sentinel nodes (melanomas, breast).

The metastases' dissemination mechanism takes advantage of the network created by the lymphatic system. Tumoral cells released by the primary tumour are initially destroyed by the macrophages present in the lymph nodes. The route taken by these cells is identical during the whole period of growth of the primary tumour. The first metastasis will install itself in the first node it meets, as soon as the macrophages are no longer capable of destroying the colonising cells. This place is called the sentinel node.

The identification and exact localisation of the sentinel node are very important as the absence of cancerous cells in this one tissue enables to confirm that the cancer has not spread beyond the primary tumour. The propagation mechanism of metastases being known for several decades, it was very quickly necessary to consider the surgical removal of part of the lymphatic system, starting from the tumour. In the case of breast cancer, it generally affects the lymphatic system irrigating the arm on the same side as the affected breast. This surgical ablation has been systematically practised since, and effectively prevents many cases from recurring. Unfortunately this technique has the drawback of many side effects and high morbidity. Surgeons have therefore considered substituting limited excision of the primary tumour as well as the sentinel node.

Original detection techniques consisted of injecting a dye around the primary tumour, specifically methylene blue, which, as it diffused, followed within minutes the same path as the cells released by this tumour. The surgeon only had to remove the most coloured area, detected during the operation. The technique is limited as particles are diffused too rapidly due to their small size, and also because of the non-visibility of the nodes located deep.

Radioactivity has recently come to the surgeon's aid. By replacing the dye with neutral particles the same size as the circulating cells, and labelling them with a short-life radioisotope, it is not only possible to follow the trace of the diffusion of tumoral cells, but also to mark the sentinel node depth.

In practice, the nuclear physician injects a suspension of nanoparticles at several points around the patient's tumour using a syringe (particles a few millionths of a millimetre in size) labelled with Technetium 99m (6 hours half-life). After a quarter of an hour, these particles will have diffused and will have started to accumulate in the sentinel node. Using a probe, a small radioactivity measuring device specially adapted for this technique, it is possible to detect the zone where this radioactivity has accumulated and, taking into account the measured value, even the node depth can be estimated, thus allowing identification.

Successive surgical ablation of the tumour and the node can be carried out the next day, once the radioactivity has decayed.

The formation of a metastasis is determined with the histopathological analysis of the sentinel node. A negative result means the cancer has not spread. A positive result indicates the need for additional examinations and regular monitoring in order to decide if more radical therapy is required.

The technique has proven to be equally effective in the treatment of melanomas, which have the particular characteristic of forming their first metastasis in sites located several tens of centimetres from the initial tumour.

Summary

Nuclear imaging makes use of radioactive substances (radionuclides combined with an organic molecule, the **vector**) which, once injected into the patient, have the particularity of distributing themselves and specifically concentrating in tissues according to the type of vector on which the radionuclides are fixed. By using gamma emitting isotopes, it will be possible to obtain a planar image of a defined zone and make a diagnosis of the evolution of the disease affecting this tissue. This method of **functional exploration** is called **scintigraphy**.

Associated to detectors revolving around the patient and powerful computing tools, it allows cross-sectional images to be obtained which, when combined together, lead to the production of three-dimensional images. This is **SPECT** technology (Single Photon Emission Computed Tomography).

Practically all tissues and organs have been the subject of the development of specific vectors with which, in most cases, Technetium 99m has been combined. For certain tissues, more specifically the brain, Iodine 123 has proven to be of particular interest.

Gated SPECT is a dynamic mode image acquisition technique coupled to the movement of the heart (or lungs), with which these disruptive factors can be ignored.

PET (Positron Emission Tomography) takes advantage of the simultaneous emission of two gamma photons in two absolutely opposite directions, requiring equipment fitted out with detectors arranged in a ring around the patient.

Finally, local monitoring of the distribution of radioactivity can be carried out with probes. This technique is more particularly applied in **the detection of sentinel nodes,** allowing a breast cancer tumour to be removed more efficiently, and considerably reducing the risk of the later formation of metastases.

CHAPTER V

PET Imaging: Positron Emission Tomography

Positron Emission Tomography (PET) uses the particular properties of beta-plus (β^+) emitters which are injected into the patient in the form of labelled products, binding specifically to the cells for which an image is required. Positron emitters have the particular characteristic of producing two gamma photons by annihilation. These radionuclides could therefore be used and detected with traditional spectrometry tools. However, these gamma photons have two additional useful properties: firstly they are emitted in two opposite directions from each other, and secondly they have the same energy of 511 keV, whatever the isotope used. This technique therefore requires a suitable imaging tool and benefits from its power and specific characteristics.

As soon as Fluorine 18's physical and chemical properties identified it as being the ideal radioisotope for this imaging technology, and when its industrial production became a real possibility, this technology experienced a surge in development, initially led by the reference labelling product, FDG. Positron emission tomography is on the verge of overtaking SPECT techniques due to its versatility and is set to soon become the new functional imaging standard.

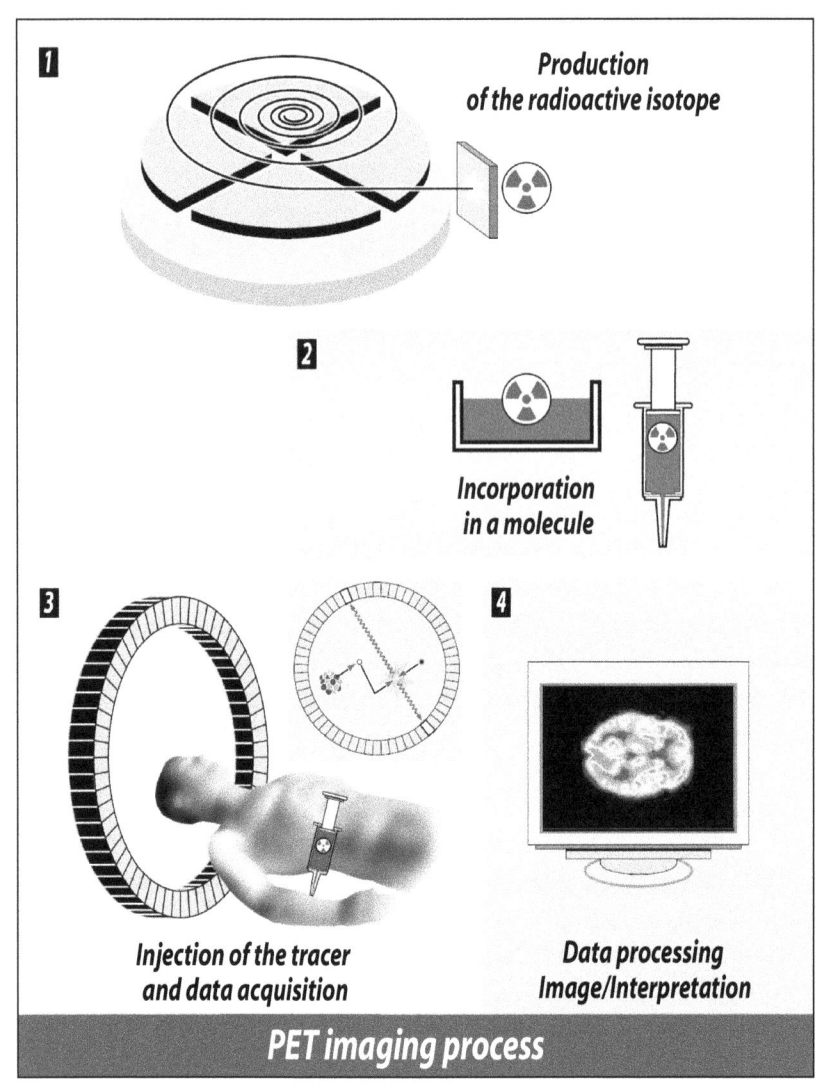

Figure 8. As a consequence of the short half-life of Fluorine 18, PET imaging requires a four-step process that has to be performed in the shortest possible time: manufacturing of Fluorine 18 in a cyclotron (1); synthesis of the FDG molecule in a radiopharmacy unit (2); transportation of the doses to the imaging centre; injection to the patient and acquisition of the image with a PET camera (3). The last step, which consists in processing data and interpreting the final images (4), is obviously independent from the previous steps.

I. The Imaging Principle

Beta-plus radiation (β⁺) is an emission of positively charged electrons, therefore of antimatter. The positrons are particularly unstable, and as soon as they meet electrons, self-destruct (are annihilated) and emit two photons which travel away from each other in strictly opposite directions with an energy of 511 keV. By placing suitable detectors on both sides of the emission site, linked with computers of sufficient power, it is possible to locate the point of origin of the collision between the electron and the positron. Imaging analysis techniques combined with recording data in successive slices (tomography) enables the creation of two-dimensional, and even three-dimensional, images. This is the Positron Emission Tomography method, or more simply PET.

Unfortunately, the positron may travel a distance of several millimetres from its ejection point before encountering an electron, due to its high energy. The final image, corresponding to the sum of the points of impact, will give the statistical distribution of these annihilation points and not the distribution of the points of origin of the positron emission. That difference of a few millimetres from the actual origin of the β⁺ signal also expresses the insurmountable resolution limit, and therefore the limit of the method's image quality. This resolution of a few millimetres is nevertheless considered as excellent due to the specific nature of the vectors used. Certain micrometastases, that is to say, metastases and tumours approx. 3 to 5 mm in diameter still remain visible due to the high level of contrast with the background noise, but their actual size cannot be evaluated. This property of imaging very small zones is not linked to the radioisotope, but to the specific characteristics of the vector on which this element is grafted.

Several β⁺ radionuclide emitters have been used, but none has succeeded in unseating Fluorine 18, a versatile element by virtue of both its radiological and its physical-chemical properties.

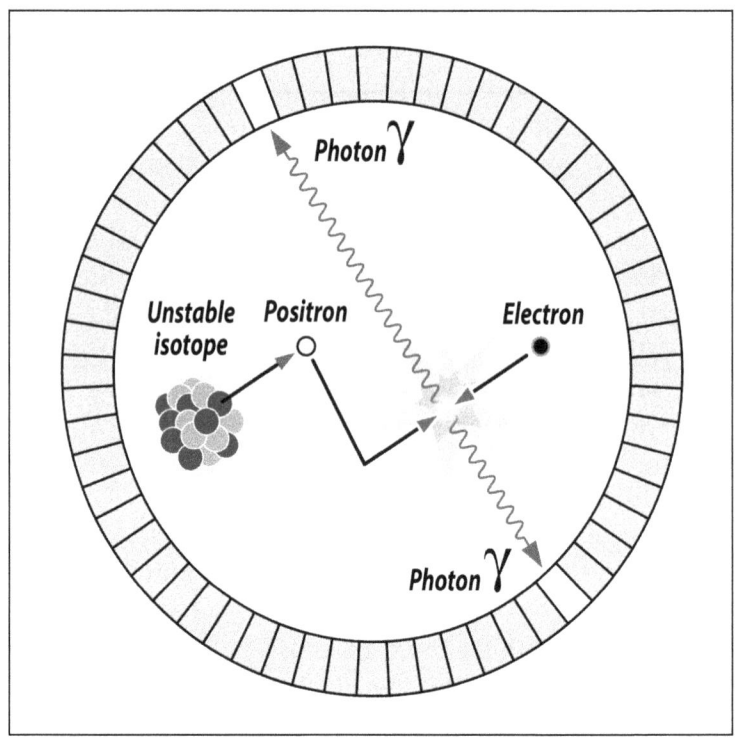

Figure 9. Scheme of the positron emission and of the detection of both gamma photons generated by the annihilation process. Detectors are placed to form a crown shape around the central source.

II. The Radiation Source

The origin of the radiation used in this technique is an atom that is very small in size, Fluorine 18, belonging to the halogen family and extremely electronegative. It has the particular characteristic of being able to form a very stable covalent (strong) bond with carbon, unlike the other halogens, and in particular unlike the extremely labile iodine atom. Due to this fact, it does not require ligands as for the cumbersome metal coordination chemistry. Because of its very low atomic volume, less than two times greater than hydrogen, fluorine can easily replace other atoms in the active molecule without radically interfering with the biological properties of these substances.

From the point of view of its physical-chemical properties, fluorine is an ideal candidate for labelling imaging molecules. Unfortunately, it has a half-life of just 108 minutes. This unusual property is both an advantage and a disadvantage. On the one hand, the short half-life promotes rapid elimination and low waste accumulation impact. The patient also remains in limited contact with the radioactivity. On the other hand, the time allocated to its manufacture, synthesis, analysis, logistics and to the imaging procedures must be adapted to this severe constraint.

Fluorine 18 is a pure β^+ emitter producing two 511 keV photons, that is to say without any harmful secondary radiation. Using a cyclotron, this radioisotope is produced by bombarding with a proton beam a target containing Oxygen 18 enriched water, a stable isotope of oxygen present at 0.2% in nature. The automated method of purification on exiting the cyclotron enables this isotope to be obtained in a form free from any radionuclidic impurity that may have been formed at the same time. The resulting aqueous solution which can be used by radiochemists contains a very dilute form of active sodium fluoride. The chemistry developed around this isotope has allowed the development of the synthesis of fluorinated organic molecules essentially by nucleophilic substitution reactions with high yields. Any general synthesis method of quantitative introduction of fluorine at low temperature without generating toxic by-products (or that are easily separable) would greatly improve this technology.

As a matter of interest, it should be remembered that other radionuclide positron emitters, in particular Carbon 11, can be used in PET on specifically labelled molecules, but none has given rise to the development of a commercial product, because of its shorter half-life (20 minutes).

III. The Labelled Product: Fludeoxyglucose

Cancerous cells are avid for glucose, and consume more of it than normal cells. Labelling glucose with a radionuclide helps distinguish between sugar consuming cells and the others. [^{18}F] fludeoxyglucose ([^{18}F] - fluoro-2-deoxy-D-glucose), better known as FDG, is a sugar, in which one of the hydroxyl groups (alcohol function) has been substituted by a Fluorine 18 element. This substitution has been carried out in such a way that during the step of ingestion by the cell, the molecule is not only still recognised by it, but also undergoes an initial metabolic transformation (phosphorylation) which prevents it from leaving this cell again. As the second phosphorylation step cannot take place at the fluorinated site as it should, the molecule remains trapped in this cell in the same way as all the other identical molecules which have been assimilated, creating an accumulation and therefore an increasingly strong radioactive signal.

The remainder of the FDG not consumed by the cells is very quickly eliminated via the urinary tract. The fluorine trapped in the cells disintegrates rapidly. According to this isotope's rapid decay rate, only less than a thousandth of the initial radioactivity dose can be detected in the body 20 hours after the injection.

FDG is injected intravenously with a dose of about 350 MBq. All the glucose consuming cells, and particularly all active cells, will trap this molecule: besides cells in the course of rapid growth and proliferation, including tumours and their metastases in particular, the brain and the heart can also be distinguished, thus opening up other possibilities for imaging. Muscle cells, which also consume glucose, will be less visible if the injection is preceded by a short period of rest on the part of the patient.

First discovered in the 70's, FDG has been widely used on patients in the US since the end of the 80's. Its first official release with a European marketing authorisation dates from November 1998, in France, while it was also used occasionally and for clinical research purposes by a few other research centres with an available cyclotron. Some other countries, Germany and Belgium in particular, had already been using this technology in hospitals for ten years. Since obtaining this marketing authorisation and enjoying increased support by several governments, a network of cyclotrons, combined with automated FDG production units, is in the course of being set up. Major funding has been released in order to equip several European countries with PET cameras and thus catch up with the United States.

Nevertheless to date, and although numerous fluorinated molecules have been synthesised, FDG remains the only substance available on the greater scale required to meet the needs of PET technology.

IV. Production and Equipment

FDG production requires mastery of the production of Fluorine 18, its purification, its use via an automated technique in the active molecule's synthesis and, finally, the quality control of the final product before distribution. All these operations must be carried out in a few hours, so as to lose only the minimum of active matter. Taking into consideration the requirement to inject the molecule into the patient as quickly as possible after its synthesis, no better solution has been found than to relocate the isotope and labelled molecule's manufacturing close to the customer sites. Production units equipped with a cyclotron are built at strategic locations which allow coverage of potential users. Cameras need to be distributed less than 3 or 4 hours away in terms of distance, independently of the means of transport used. Beyond this time period, the amount of matter lost becomes too great, and the cost of doses as well as that of transport increase proportionally.

Every hour, a little more than one patient can be evaluated with each camera. Newer camera models can scan up to three patients in that same time. Each production unit, with its cyclotron, has the status of a pharmaceutical establishment and must follow current pharmaceutical manufacturing practices. They are managed by a head of compliance pharmacist who guarantees the delivered products' pharmaceutical quality.

The number of cyclotrons is set to stabilise very quickly, given the funding required for the devices and their infrastructure, at the level of 4 million euros. The number of patients to benefit from this technology in the future will depend essentially on the number of cameras available. The new hospital equipment plans, which allows major centres to make investments in this field, actually fixed the average rate at one camera per million inhabitants. As an example, France, the most advanced European country in terms of industrial centres first approved 60 PET cameras that could be supplied by about 10 cyclotrons. Very recently (2005), this figure was revised upwards and 80 cameras should be deployed in this country by 2008. It is very probable that this figure will grow further again, in view of the new applications this technique offers in oncology, haematology, cardiology and neurology. The other European countries, which are lagging behind excepted for Germany and Belgium, will certainly follow suit and be similarly equipped by 2010.

Given the technology's evolution, the new equipments installed are mainly hybrid systems combining Positron Emission Tomography with Computerized Tomography X-ray cameras. These tools, also known as PET/CT, were invented in 2000 and have been marketed since 2002; they allow superimposed images from both procedures to be obtained, and thus pinpoint the observed elements with much greater precision.

V. Applications in Cancerology

The official indications for FDG as described in the European marketing authorisation dossiers cover the following areas: FDG can be used in the characterisation of isolated pulmonary nodules and the diagnosis of metastatic cervical adenopathy of unknown origin. It serves for the evaluation of primary pulmonary cancers, including the detection of remote lung metastases, the evaluation of the head and neck tumours including assistance in guided biopsy, as well as recurrent colorectal cancer tumours, lymphoma and malignant melanoma. As a means of monitoring therapeutic response, it is used for head and neck cancers and malignant lymphoma. Finally, in case of reasonable suspicion it is used to detect recurrences of head and neck cancers, primary lung cancer, colorectal cancer, malignant lymphoma and malignant melanoma.

In reality, due to its universal nature, FDG is capable of detecting the appearance of tumoral or metastatic cancerous cells for almost any so-called solid cancer. Nevertheless, a demonstration of its effectiveness has not yet been carried out in all cases and other, less costly techniques may provide equivalent information. In particular, it may well serve for detecting breast cancer, and even help its prevention, but other effective detection methods can confirm diagnosis faster.

FDG becomes useful when a tumour is suspected, or when a clear diagnosis cannot be obtained with more traditional methods due to a lack of suitable tumour markers or because the zone being considered is difficult to access.

Beyond FDG, numerous vectors developed on the basis of biological mechanisms and labelled with Fluorine 18 have found specific applications in oncology. We will only mention the molecules of simple structures such as FNa (sodium fluoride), FLT (fluoro deoxy-L-thymidine), FDHT (fluoro 5-alpha dihydro testosterone), FETNIM (fluoro erythro nitroimidazole) or FMT (fluoro alphamethyl tyrosine) which are used for the specific detection of bone metastases, breast cancer, cancer of the prostate, the larynx or melanomas, respectively. Unfortunately, and despite their diagnosis potential, the use of these molecules remains mostly restricted to clinical research, due to their high cost, their low return on investment and the complexity of their transformation into medicines approved by the European medicine agencies.

VI. Applications beyond Oncology

The heart and the brain consume great amounts of glucose, and are therefore potential candidates for imaging using FDG. However, these organs show a higher background noise, which is often a hindrance.

FDG's cardiac applications are challenged by more traditional, and above all less costly, techniques such as scintigraphy using Technetium 99m or Thallium 201. Nevertheless, these methods also have their limits, and PET could possibly be of greater interest if FDG was confirmed to be useful in measuring myocardial viability, otherwise very difficult to determine. The use of vectors specific to cardiovascular mechanisms will open other, more promising, avenues. PET has made its greatest contribution in the diagnosis of neurodegenerative diseases. Almost all the active molecules in neuronal cells can be labelled with fluorine. Unlike technetium and other metals requiring chelating groups in order to be attached to the vectors, the simple chemistry of fluorine prevents any additional biological constraints from occurring and allows the molecules to pass the blain-blood barrier. All fluorinated derivatives retain this property quasi identically to the non-labelled vector, which enables them to access the neuronal receptors at particularly high concentrations. In this case, FDG is only of limited interest, as it is not sufficiently specific. Other molecules have been labelled with Fluorine 18, thus allowing particular pathologies to be studied, and even brain behaviour to be observed. These vectors are called dopamines, serotonins or benzodiazepines.

VII. Positron Emitters Evolution

Among the available isotopes, fluorine is not the only one worthy of interest, but to this day remains the only one that led to a commercially available drug. Among other β^+ emitters produced in a cyclotron, let us mention Carbon 11 (20 minutes half-life), available in the form of carbon dioxide or methyl iodide which could easily be integrated into an organic molecular structure were its half-life not so limited. Less than half an hour to carry out a synthesis, followed by purification and quality testing, is a very constraining time for the chemist. In fact very few molecules are suitable for Carbon 11 incorporation, and all those developed to date will presumably remain restricted to clinical trials. Oxygen 15 is even more unusual by virtue of its half-life (2 minutes) but is nevertheless used in the form of labelled water, purified and injected directly as it comes out from the cyclotron. This technique allows physicians to obtain information on the irrigation of certain tissues and fluid exchanges.

Among β⁺ isotope emitters with a longer half-life, Gallium 68 may be of interest as it is manufactured in a generator from the decay of Germanium 68; therefore, it does not require the heavy installation of a cyclotron. Its metal character obliges chemists to develop more complex organic structures. To this day, no product based on Gallium 68 has succeeded in breaking through. Radiochemists are on the lookout for new β⁺ isotope emitters, but essentially come up against production and physical-chemical properties problems. Fluorine will probably remain the best β⁺ emitter on the market for decades to come.

Summary

Positron emitters (β⁺), generating positive electrons (anti-electrons), produce two photons as a consequence of each collision with an electron (**annihilation**). These two photons, which both have the same constant energy, are emitted in two totally opposite directions. Detectors are placed on either side of the emission point to accurately locate the origin of the source. In practice, several detectors are placed in a ring around the origin of emission.

Among the positron emitters that can be used in nuclear medicine, **Fluorine 18**, with a half-life of about two hours, has found an inestimable advantage in diagnosis medical imaging. It is through the use of the fluorinated glucose derivative, **FDG** (fludeoxyglucose) that **PET** (**Positron Emission Tomography**) has demonstrated its effectiveness in oncology.

Nevertheless, implementing this technology requires the creation of structures (cyclotrons, synthesis laboratories, etc.), located and adapted to its short half-life. Furthermore, the development of the technology will depend on the investment in cameras made in the territory being considered.

Other fluorinated products are currently under evaluation and will soon allow the evolution of a pathology in the course of its treatment to be diagnosed, and then monitored, in fields as diverse as oncology, haematology, cardiology and neurology.

CHAPTER

Therapeutic Methods

The ionising effect of radiation can quite naturally be useful in therapy by destroying supernumerary or abnormal cells. When a radionuclide is attached to a substance which aims to concentrate in a certain type of cell, it can destroy these same cells. This is the simplified principle of targeted radiotherapy (radioactive source attached to a drug or vector).

The property of a radiopharmaceutical that causes it to concentrate in a particular tissue or organ, in other words its specificity, is uniquely due to the property of the vector, that is to say the molecule to which the radionuclide is attached. Iodine, in the form of iodide salt, is an exception because this element, whatever its saline form may be, spontaneously accumulates in the thyroid tissues whether these are cancerous or not, and therefore does not need to be combined with a vector. The vector's specificity is shown by its tissular radioisotopic distribution. The image obtained from the gamma radiation of a gamma emitting radionuclide attached to a vector is an exact picture of this molecule's distribution in the patient's tissues. Replacing the gamma emitting isotope of this diagnosis radiopharmaceutical with a beta or alpha emitting isotope will not change the labelled molecule's tissular distribution. Contrary to gamma emitting radionuclides, ionising-type irradiation will affect infiltrated cells in a radically different manner. Hence this radionuclide substitution leads to the transformation of an imaging radiopharmaceutical into a therapy radiopharmaceutical, with a local cellular destruction effect. In theory, this principle seems to be very simple; in practice, it is only effective if the substitution of one isotope for another on the same vector can be done without chemical and biological modification of the latter.

The technology can be applied well beyond metabolic mechanisms and any biologically "active" molecule present is potentially worth labelling, whether one wishes to study hormonal or enzyme mechanisms, the interaction of ligands and receptors, substance transport mechanisms, or even cellular reproduction. It would be worth replacing the frequently used term metabolic radiotherapy by internal radiotherapy or, more precisely, targeted radiotherapy.

I. Metabolic Radiotherapy

Radionuclides which intervene in the metabolism of a cell, *i.e.* in the mechanism of its biological functioning, can be used for therapeutic purposes since their damaging effect will operate within the actual cell itself. This is basic metabolic radiotherapy.

Iodine naturally and specifically accumulates in the thyroid; treating patients with radioactive iodine has proven to be unexpectedly effective in all of this tissue's affections. Iodine was the first radionuclide to be administered to man on a large scale with a therapeutic goal in view, and has given no reason to doubt its effectiveness since the 40's. Given orally and in its simple sodium salt form, Iodine 131 is used for treating benign thyroid diseases such as Grave's disease (diffuse goitres, an immune system disorder which disrupt regulation of the secretion of thyroid hormones), toxic forms of goitres, hyperthyroidism and in the elimination of benign thyroid nodules or post surgical residuals. It is also used in therapy for treating malignant conditions (thyroid cancer), including metastases, as long as the tumour is able to accumulate this element. Iodine 131, an emitter of both γ and β^-, is also used at a weaker dose for imaging the thyroid and any of its possible metastases, while Iodine 123 remains limited, in this pathology, to its current use in imaging.

Today, radioactive iodine remains the favoured treatment for this type of cancer, and was for a long time the only really effective

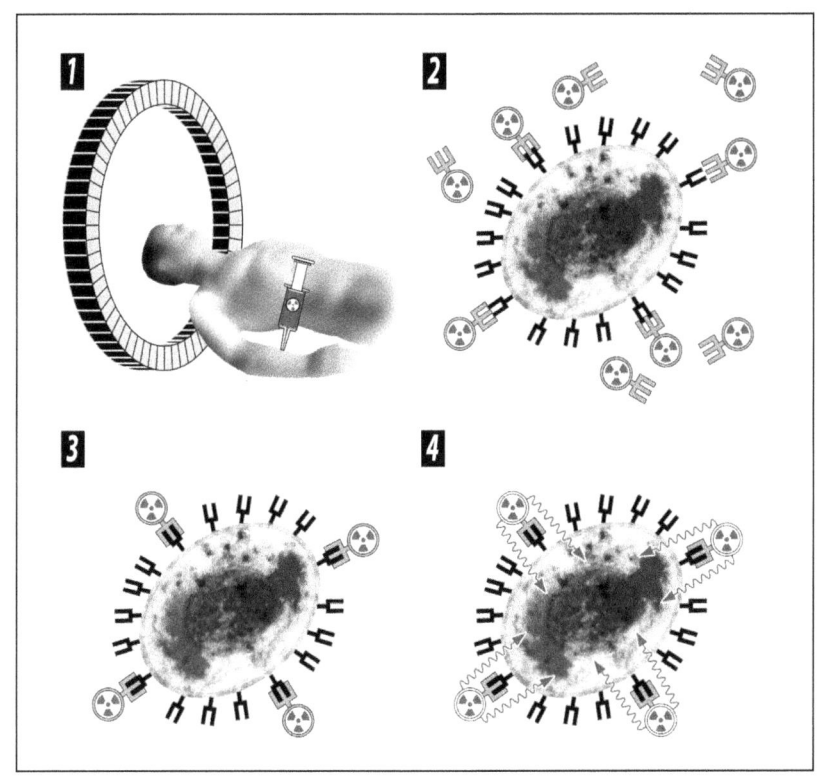

Figure 10. Metabolic radiotherapy principle: the radioactive substance linked to a vector (drug) is injected to the patient (1). There are large numbers of specific receptors on the surface of the tumour cell, which the vector is able to recognise. The vector binds to the cell receptors (2). As unbound molecules are naturally eliminated from the body (3), the radionuclide slowly disintegrates and destroys any cell present in its immediate surroundings by emitting beta-minus particles.

metabolic radiotherapy tool. There is no other radioisotope-tissue pair displaying such a strong bond and such evident specificity.

Structural analogies between the components of bones and certain radioactive atoms could explain the mechanism underlying the action of substances used in the palliative treatment of pain in patients suffering from bone metastases.

The formation of these metastases, particularly in the evolution of breast or prostate cancers, can be extremely painful, and is today considered as the beginning of the cancer's terminal phase.

Treatment becomes extremely radical and healing, even remission, are very unlikely. High doses of morphine are usually given as a pain killer. Among other methods, products such as Strontium 89, Rhenium 186 and Samarium 153 in the form of salts or complexes have been developed to provide at least partial relief from the intense pain which is generated by the metastases and develops on contact with bone tissue.

It seems that their analogy with bone constituents enables them to insert themselves between the metastasis and the bone tissue, thus significantly reducing pain, and even enabling patients to continue their anti-cancer treatment as outpatients. In some cases, recent observations have shown a reduction, and sometimes a halt, in the disease's evolution. Researchers even think that in view of the specificity of these agents, they may, at higher doses, promote the disease's regression. It is too soon to say if these radiopharmaceuticals are truly effective from a therapeutic angle, but studies are in progress with a view to demonstrate either their effectiveness or the lack of it. Initial results are not expected before 2007 or 2008.

II. Local Radiotherapy

Local radiotherapy consists of injecting a radioactive substance into a specific delineated natural zone or cavity with a therapeutic effect that will in consequence be limited to this region.

The treatment of rheumatoid arthritis is one of the few non-oncological pathologies that can benefit from this internal radiotherapy technique.

Rheumatoid arthritis is a phenomenon that is allergic in origin, and which leads to the abnormal proliferation of synovial cells filling the cavity between the two segments and promoting the destruction of the cartilage cells. This auto-immune disease is treated in the first place with anti-inflammatory substances to halt its progression. At a later stage, physicians use drugs often analogous with anticancer treatments (cyclophosphamide, methotrexate, TNF-alpha) to reduce the abnormal growth of synovial and cartilage cells. A treatment possibility is the destruction of supernumerary cells with drugs or through surgery, coupled with palliative pain treatment (corticosteroids).

The substances used in these therapies often lead to undesirable side effects. Radio synoviorthesis may be recommended when the doctor finds that traditional therapies have failed and when the disease becomes more acute and spreads to other joints. This internal radiotherapy technique, known since the end of the 60's, is based on the destructive effect of β^- radiation. A radionuclide, chosen on the basis of specific physico-chemical and radiological criteria, is injected into the joint cavity containing the synovial fluid. Along with its half-life, the beta emitter's energy is of utmost importance as it determines the average effective distance. Three different radionuclides are used for this purpose: Erbium 169 (9.5 days half-life, 3 mm average effective distance in the tissue, with a 10 mm maximum), the least energetic, is prescribed for hand joints. Rhenium 186 (3.7 days half-life, 12 mm average distance, 37 mm maximum) is preferred for wrists, elbows, shoulders and hips, while Yttrium 90 (2.7 days half-life, 36 mm average distance, 110 mm maximum) is used for injection into the knees.

In order to prevent the isotope from spreading outside the inter-joint space, it is trapped in neutral microparticles and the product is injected in the form of an aqueous suspension.

The results show a spectacular recovery of the capacity to use these joints, a pain reduction in most cases, with an improvement still clearly marked five years after the first injection, without the need for additional treatment. The use of these products is limited by the lack of specialists in this technique, the need for in-patient treatment and in some countries, the absence of reimbursement.

III. Radioimmunotherapy

The combination of antibodies and radionuclides leads to the formation of highly specific drugs which allow either the imaging or treatment of certain types of pathology due to their involvement in the natural immune mechanisms. This is the radioimmunodiagnosis technique that can be combined with radioimmunotherapy.

In a process of self-defence against external chemical or biological attacks (antigens), the human body is able to develop specific substances (antibodies) that can destroy these foreign invaders. This is the basis of our immunological system's functioning that produces new antibodies each time new external molecules or foreign cells, which are not recognised as being fully part of the body, enter the cells or the tissues. In fact, antibodies recognise a molecular structure of these external entities which is called an epitope. Abnormal cells and precursors of cancerous cells are also eliminated through this mechanism. Unfortunately at a certain stage some of them succeed in colonising a tissue and give way to a tumour. Growing cancer tumours are not considered as foreign entities anymore, and therefore are not destroyed by the immunological system. However, they show some specific epitopes in larger amounts than surrounding normal cells, and as a result they can be recognised by matching antibodies.

Antibodies can be used to fight diseases and, as they are specific to an antigen, are ideal vectors for targeting tissue as long as they are also capable of crossing the biological barriers. Numerous antibodies that selectively target tumoral antigens have been developed over the course of the last two decades. They are used in the context of immunotherapy treatments. The pathogen potential of these substances is increased by grafting a β^- emitting radiotoxic substance onto them in order to transform them into a radioimmunotherapy product. The use of γ radionuclide emitters in the place of β^- emitters enables the distribution of these antibodies to be displayed and gives access to radioimmunodiagnosis or immunoscintigraphy tools.

Numerous technical barriers have had to be crossed, because these products constitute the pinnacle of complexity of a drug: they are subject to the constraints of pharmaceutical quality manufacturing, to the obligations linked to the use of biological material of animal or human origin, and to compliance with all the safety aspects linked to the use of nuclear materials. Polyclonal antibodies (originating from several stem cells) of animal origin have given way to humanized chimeric antibodies (the animal origin part must not be recognised as foreign by the human immunological system), and practice is increasingly moving towards the production of modified monoclonal antibodies of human origin (single stem cell source). In this way abnormal generation of human anti-mouse antibody (HAMA) is avoided. The radioactive labelling methods for these molecules do not target a precise site on the antibody, but ensure that the radioisotope is fixed in a part of the macromolecule that does not interfere with the antigen recognition zone. Today, labelling chemistry has advanced to such a degree that, in practice, it is possible to pair almost any type of isotope with any antibody without losing the biological properties of this antibody.

The trial and error period looking for the ideal radioisotope seems to be coming to an end and several radionuclides such as Iodine 131 and Yttrium 90 stand out for reasons essentially linked to production ease, to chemistry, safety and dosimetry issues and above all to their physical properties, such as half-life and energy. A few other isotopes are still being put forward and studied by researchers, but this is presently limited to Holmium 166, Lutetium 177 and Rhenium 188.

On this basis, a number of antibodies have already been the subject of evaluation: by way of example, we would mention anti-CD20, anti-CD21, anti-CD22, anti-CD37, antiferritin and anti-HLA-DR. These macromolecules have principally been labelled with Iodine 131 or Yttrium 90. Labelling with Technetium 99m or Indium 111 has enabled the distribution of certain of these substances in the organism to be studied. Among radiolabelled antibodies, the use of which is, on the face of it, limited to imaging and which are being developed or marketed, we would mention arcitumomab labelled with Technetium 99m for imaging colorectal cancer, capromab labelled with Indium 111 for imaging prostate cancer, antigranulocyte MAb 250/183 labelled with Technetium 99m for imaging infections and inflammations and satumomab labelled with Indium 111 used in the immuno-scintigraphy of colorectal and ovarian cancer.

The first two anti-CD20 antibodies, tositumomab and rituximab, labelled with Iodine 131 and Yttrium 90 respectively, were made available on the market in 2003 for the treatment of patients suffering from non-Hodgkin's lymphoma and resistant to traditional chemotherapy, including immunotherapy. Clinical studies in progress should demonstrate that these products are even more effective when administered as the first intention treatment.

Radioimmunotherapy and Lymphomas

Antibodies are substances with very complex structures. They are produced by an organism, and are intended to initiate a reaction in order to destroy a foreign body. Antibodies are capable of recognising a fragment of an infectious agent or a molecule that is foreign to the organism, called an antigen. Each antibody has its specific attachment sites. Our organism is capable of producing new antibodies each time it comes in contact with an undesirable substance. As and when contact with external elements occurs, our organism builds natural barriers that contribute to our internal defence mechanism called the immune system. The different antibodies present in our organism number several million and are permanently on guard. In the context of an external threat, our immune system is capable of producing millions of copies of the antibody specific to this aggressor in order to fight it. Obviously, it can only do so if it has already been in contact with this aggressor or this disease before. An allergic reaction is an excessive reaction on the part of the organism to an external intruder for which the body is usually prepared, that is to say for which the organism has already built its antibody. Vaccination is a form of prevention that works by introducing these elements into the organism prior to any contact with the disease. This very simplified explanation helps us understand that, normally, a disease only occurs in the case of a first contact with the infectious agent, or else when the organism is not capable of reproducing its own defence agents quickly enough.

As a tumour is made up of abnormal cells, one might think that the organism would be capable of developing antibodies likely to initiate its destruction. Unfortunately, tumoral cells originate from the individual's normal cells and retain certain characteristics of these healthy cells. As a consequence, they are not recognised as foreign cells by the organism and are not therefore fought against. On the other hand, antigens present in the tumoral cells can be identified which do not exist in healthy cells, or at least not to the same extent. Biologists have succeeded in developing antibodies that specifically target these antigens and are therefore able to attach themselves mostly to the tumoral cells. All that remains is to transform these new vectors into media for a toxic substance capable of destroying the tumoral cell. This is the basis of immunotherapy. If this toxic substance is a gamma emitting radioisotope, a radioimmunodiagnosis or immunoscintigraphic product will have been built that will serve to display the distribution of the labelled antibodies in the tumour. By using a beta-minus or alpha emitting radioisotope, the radioimmunotherapy product will serve to destroy this same cell.

The concept seems extremely attractive. A number of technical barriers have had to be crossed. The initial "artificial" antibodies were formed from tissue taken from mice or rats. They had the disadvantage of being, in part, rejected by the human organism which began making anti-mouse antibodies. In a second approach, biologists succeeded in creating mixed mouse-human antibodies, called chimeric antibodies, and it is only recently that they have been able to construct human antibodies easily.

In between times, radiochemists also developed methods of radioactive labelling of these molecules. As these molecules are extremely large, it was pointless to want to develop a method targeting a precise function of the molecule. Rather, chemists made sure they could attach a radionuclide to a part of the macromolecule that does not interfere with the zone that serves for recognition of the antigen. The chemistry is now developed, and it is possible to pair (almost) any type of isotope, including alpha isotopes with (almost) any antibody.

The application of these principles is described in the example of treating non-Hodgkin's lymphoma.

Non-Hodgkin's Lymphoma

Non-Hodgkin's lymphoma is a disease that is expressed by the formation of tumours in the lymphatic system. Twenty or so distinct lymphoma sub-types have been noted. The difference from Hodgkin's disease can only be seen by analysing the cells under a microscope. Hodgkin's disease is characterised by the presence of a very specific type of cell called the Reed-Sternberg cell.

Generally speaking, the first signs of the disease are expressed by a painless swelling of a lymph node. As these nodes are situated in all parts of the body, the first discomfort depends on the location of this abnormal cell development. In the case of the digestive system, the initial consequences are expressed by nausea and vomiting or abdominal pain, while a node in the stomach may affect the breathing capacity. Headaches, sometimes combined with visual problems, can indicate a brain tumour, and when the bone marrow is affected, anaemia will be the first consequence. The disease may cause a hyperactive immune response. As the symptoms are often exacerbated when the body is fighting an external threat such as an infection, this response is expressed by fever, tiredness and weight loss. The immune mechanism behind this phenomena is similar to that causing itching, for example. It is considered to be a fatal disease, however more than half of sufferers have a survival expectancy of more than 5 years, and this period can only increase. The disease affects all age groups, but half the cases are people of more than 60 years of age. There has clearly been an increase of this disease over the last 20 years, for which there is no explanation. Neither can its origin be explained. It became known to the general public via famous cases like those of Jacqueline Kennedy Onassis and Shah Mohammed Reza Pahlevi.

Treating Non-Hodgkin's Lymphoma

Major efforts have been devoted to the development of a treatment for non-Hodgkin's lymphoma, due to the limited response rate to current treatments and the high rate of side effects associated with chemotherapy and external radiotherapy methods. Although the development, over the last 40 years, of these methods allows an extension of this disease's remission time, it hasn't yet shown any significant improvement in the survival rate.

From Indium to Yttrium

Several monoclonal antibodies have been developed and marketed that are of animal or chimeric origin, that is to say mixed, and therefore formed of molecules coming both from humans and mice, and directed against the CD20 antigen present on the surface of malignant lymphocytic B-cells typical of the disease to be treated. These antibodies have a therapeutic activity by themselves, as they block the functioning of these malignant cells by preventing them from reproducing. Unfortunately, due to their mechanism, they cannot claim to destroy these cancerous cells completely and the treatment, though effective, only manages to slow the evolution of the disease down.

Knowing that these antibodies target the cells to be destroyed, it becomes obvious that by combining them with a toxic substance, an effect on this target could be hoped for, and possibly one reaching abnormal cells within the immediate environment.

In order to verify whether the antibody is actually distributed to these malignant cells and to them only, a radioactive atom is grafted onto one of these antibodies, specifically Iodine 123 or Indium 111. Labelling is obviously carried out in such a way so as not to disturb the interaction between the antibody and the antigen targeted. The two radioisotopes are gamma emitters, and the images obtained after injecting the labelled products helps confirm the initial hypothesis. At the same time, these new labelled molecules become specific imaging agents for the disease.

Although Iodine 123 and Indium 111 are both Auger electron emitters, which could be used in therapy, the concentration on the surface of the B cells is too weak to expect it to be effective in any way. It has been preferable therefore, to replace these two isotopes with Iodine 131 and Yttrium 90 respectively, both high energy beta emitters. New therapy radiopharmaceuticals were born. The two substances are providing convincing results, each with its advantages and drawbacks. Thanks to this work, the yttrium and iodine labelled substances were put on the US market in 2002, and they have been available in Europe since 2004.

Taking into account the little opportunity to take stock of this type of therapy, miracles must not yet be expected. It is preferable to advance in small steps and demonstrate an improvement in patients, stage by stage. Thus these molecules have only been authorised for certain categories of patients, suffering from a non-Hodgkin's type of lymphoma, and at a very precise dose of radionuclide. In particular, treatment is only administered to people for whom previous chemotherapy, including immunotherapy using the same antibody, has failed.

Despite this limitation, these products have given positive results, demonstrating their therapeutic value as compared with existing therapies, one of the first new victories for nuclear medicine.

Studies are in progress, firstly to demonstrate that the treatment is at least as, if not more, effective if given earlier, and secondly to increase the dose in order to obtain an even greater rate of response. These studies, which need to take place on a wide scale, will still require several years of work before authorisation is given for an extension to the indication. Nevertheless, these examples show the significance and the effectiveness of this method. A number of radiolabelled antibodies are being evaluated with a view to treating various forms of cancer, and could be marketed within the next five to ten years.

IV. Targeted Radiotherapy

Targeted radiotherapy combines a therapeutic radionuclide, mainly a β^- emitter, with a smaller organic molecule serving as the vector and which will target the tumour cell.

In order to get round the biological stability of antibodies and the difficulty of producing them, some teams have started to label smaller sized molecules. After trials with immunoglobulin fragments, biologists and chemists have directed their interest in synthetic peptides. In fact, all biologically active molecules recognised by a receptor can serve as a vector. On the other hand, the specificity for a given target becomes of first importance, and the idea of working with antibodies started from the principle that the larger the molecule, the greater the number of recognition sites and the greater the chance of it being recognised by a single type of cell only. This theory is no longer put forward since it has been possible to show that a limited number of recognition sites participate in the antigen-antibody interaction, and that in this case one of the parts can be reduced to a very small structure that can be reconstructed by combining amino acids in an appropriate fashion. A new avenue opens for obtaining radio-labelled molecules with this particular chemistry, and peptides form another class of vector from which a great deal is expected.

In the peptide class, certain products are a few lengths ahead. Somatostatin analogues (*i.e.* octreotide, pentreotide, lantreotide and depreotide which are labelled with Iodine 123, Technetium 99m or Indium 111) provide specific images of various cancers' tumours and metastases due to their particular affinity for over-expressed receptors in these cells. The substitution of these imaging radionuclides by therapeutic radionuclides (Rhenium 188, Yttrium 90 or Iodine 131) allows the use of these same molecules to be considered in the treatment of patients. Initial clinical studies show very promising results.

V. Alphatherapy and Alpha-immunotherapy

Alpha emitters have a powerful destructive capacity due to the large size of the ejected particle. This size also limits the zone of interaction with neighbouring cells. Several tenths of a millimetre of matter are sufficient to stop this radiation. Therefore, an alpha particle has the ideal profile for use in cancerous cell destruction. Combined with the techniques referred to previously, one can speak of alphatherapy and alpha-immunotherapy.

Unfortunately, detecting alpha radiation is much more difficult than detecting beta or gamma emitters, and becomes impossible once the substance is absorbed or ingested. As a consequence, accidental contamination will also be more difficult to localise and this limitation will greatly influence its production, transport and implementation in the medical world.

Also, due to their aggressive nature with regard to all cells whether healthy or sick, alpha emitters can only really be applied to patients using extremely selective vectors which are rapidly eliminated from the body when they do not bond with their target. In this case, the effects of alpha emitters concentrate in the circulatory system (veins, arteries) and the organs involved in metabolic functions (liver, kidneys, bladder). This is unlike beta emitters which, due to their greater energy, destroy as many cells, but from a greater distance therefore in a more dilute way inside the body. The notion of an effective half-life, which gives information about the time period during which the isotope retains its destructive character in the targeted organ and which is connected to both the isotope's half-life and the biological half-life, takes on renewed importance.

Finally, alpha emitters also suffer from greater negative publicity when compared to other radioisotopes because their only applications known to the general public are essentially military.

However, if the specificity of vectors could really be improved, alpha emitters would be ideal isotopes for use in anti-cancer therapy. Researchers are continuing to explore this avenue, trying to get round the above-described constraints.

The isotopes used in metabolic therapy have very short half-lives, and even extremely short compared with the better known alpha emitters like uranium or plutonium. In fact, only three isotopes have so far demonstrated their use and potential for application in research laboratories: Actinium 225 (half-life of 10.0 days) which has the particular characteristic of producing four alpha particles during its decay, Bismuth 213 (half-life of 45 minutes) and Astate 211 (half-life of 7.2 hours).

As surprising as it may seem, alpha radiation was the first to be applied in nuclear medicine. Radium salts were also used for therapy. To this day, there is only one alpha emitting product on the market: Radium 224 Chloride (half-life of 3.62 days), used in the treatment of ankylosing spondylarthritis (Bechterev's disease). This product is only available in a limited number of countries, such as Germany.

Grafting an alpha emitting radioisotope onto a vector directed against a biological target was first attempted ten years ago with Bismuth 213. Tests in man with anti-CD33 type antibodies or labelled tenascin are in progress. Definite effectiveness has been noted, but research is progressing slowly as it follows results from a successive increase in applied doses. The maximum dose tolerated, *i.e.* before side effects appear, has not yet been reached, indicating that further improvements in patients' state of health are expected. Initial envisaged indications relate to gliomas (brain cancer) and lymphomas.

Actinium 225 trapped in a molecular cage, itself attached to an antibody or peptide, will play the same role as the beta emitters described above. Nevertheless, due to the fact that it can release four alpha particles, one after the other, its fire power is four times as great. The Americans, who developed this technology, named it the "nanogenerator": a form which, once located on its target, is capable of generating several radioactive particles at a molecular scale. *In vitro* these molecules have already demonstrated their effectiveness against various human cancer cells (leukaemia, lymphoma, breast, ovaries and prostate). Initial clinical trials have started in patients with a glioma (brain cancer), non-Hodgkin's lymphoma and acute myelogenous leukaemia.

In France, more precisely in Nantes, a major project to install a 70 MeV cyclotron is under way. It should be operational by 2009 and will be used for research and development. This tool will be twice as powerful as the largest Europe-based cyclotron dedicated to nuclear medicine, and will allow alpha radionuclide emitters to be produced in much greater quantities.

THERAPEUTIC METHODS

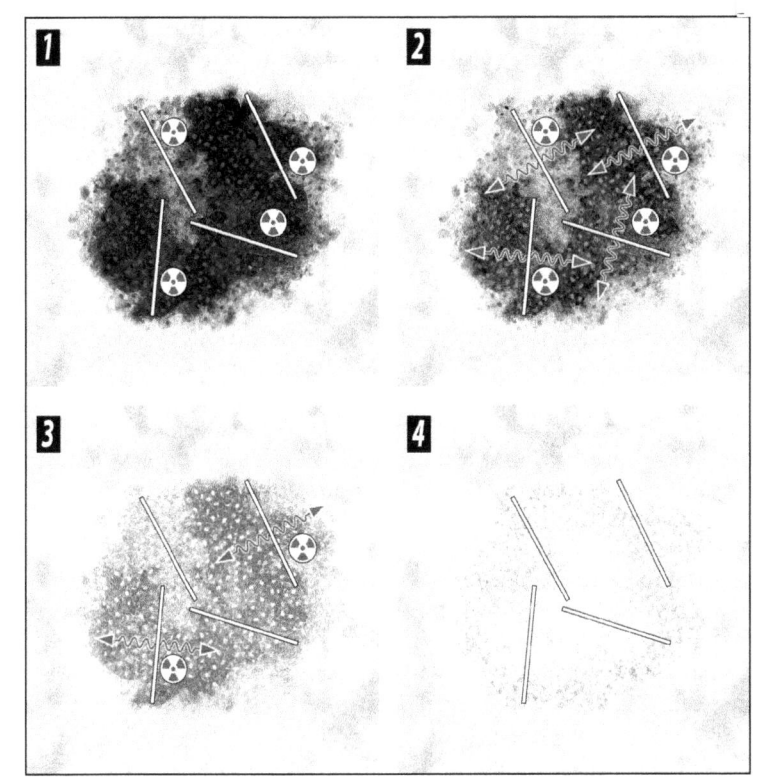

Figure 11. Brachytherapy: solid radioactive units (seeds, wires…) represented here in the shape of long rods are implanted in the tumoral mass. Radioactivity then slowly destroys the surrounding (*i.e.* tumoral) cells. The metallic implants will either definitely remain on site (prostate) or be removed later, depending on their size.

Brachytherapy and Prostate Cancer

Brachytherapy (from the Greek *brakhus*, meaning short, in other words therapy at a short distance), is expressed in the field of radiotherapy by contact treatment between the irradiating source and the organ to be treated. The French prefer to use the term "curietherapy". In practice, it's a question of simply implanting an irradiating source in the tumour to be destroyed, or in a cavity, for a predetermined period of time. Initially, curietherapy was limited to the treatment of cervix cancers. This still is the most frequent application in many developing countries where there is no access to recent therapies. These radioactive implants, initially labelled with radium, and now with iridium, caesium, strontium or iodine can take on any desired form. In the case of breast cancer, curved wires are used which remain in contact with the tumour in the breast for several hours or days. Brain tumours can be treated, in addition to surgery which needs to be the least disfiguring possible, using radioactive balls which are temporarily placed in this new cavity.

However, all these applications remain very limited because they are difficult to implement (they require special environment and equipment), are not well-known and are non-specific (results depend on the operator's experience).

Meanwhile, there exists a field in which brachytherapy has found its niche, namely in the treatment of prostate cancers. Prostate cancers develop very slowly, and affect a large part of the male population over the age of 75. Taking the concerned age group into consideration, mortality linked to other pathologies still remains predominant. It would seem that more than 80% of men who live beyond the age of 80 have the disease without being affected by it, and without being even conscious of the fact. Unfortunately, this disease also affects men under 50, who hope to be able to find a non-traumatic way of healing in medicine. As with all other cancers, the chances of healing are so much the greater when the disease is detected early on. It is estimated, nevertheless, that it takes 15 years for the disease to develop metastases, allowing patients and physicians some time to choose and intervene.

As with breast cancer, surgery remains the most common solution. It nevertheless does have a number of disadvantages, some of which are painful, such as urine retention and sexual impotence problems. For surgery to remain effective, it is necessary to remove the entire tumour contained in the prostate, and therefore several millimetres of healthy tissue as well. Frequently, surgeons are obliged to cut into muscle tissue used in urinary or erectile functions. This loss of functional capacity is no longer considered as being insignificant for people who are otherwise well, despite their age, and is even more difficult to accept by younger patients.

Brachytherapy comes to the aid of these patients by offering an alternative to surgery in the form of radioactive metal implants which are placed and confined within the tumour. About a hundred of these small pieces, less than a millimetre long (seeds), are implanted directly into the tumoral tissue under local anaesthetic. Their distribution in the tumour is arranged very precisely, in order to obtain uniform radiation of the cancerous tissues by the Iodine 125 or Palladium 103 which they contain.

Several tens of thousands of American patients have already benefited from this technology, which has become routine in the United States over the last

10 years. The emphasis is placed on its advantages in everyday life: a very short stay in hospital and the absence of side effects. The metal implants, which lose their radioactivity after a few months, remain in place for life. They do not cause any problems, unless the patient dies in the weeks following the operation, as cremation is rendered impossible by the fact that the radionuclide could be released into the atmosphere. For this last reason, the use of this technology is still under discussion in Japan.

It must however be mentioned that this technique requires extensive technical support. The operation can only be carried out by a trained and competent team. The urologist or oncologist who first discovered the cancer may recommend this procedure. It is obviously carried out in the presence of an anaesthetist, but it also calls for a nuclear physician or a radiotherapist who are, depending on the country, the only ones authorised to handle radioactive sources. Furthermore, as placing the seeds can only be correctly carried out by displaying them with an ultrasound device, the presence of a specialist in this technology is essential. Finally, special equipment, which requires a suitably large budget, is essential, and this is for the moment taken care of by the manufacturer supplying the implants.

VI. Neutron Capture Therapy

Neutrons, trapped by certain atoms such as boron, transform them into other entities which are themselves radioactive. They generate ionising radiation and therefore have a therapeutic power. By first concentrating "cold" boron in the abnormal cells, then irradiating these cells with a neutron beam, an internal radiotherapy technique is indirectly implemented. This is called neutron capture therapy.

On December 19th, 2002, the French Press Agency (AFP) announced a world first, by an Italian team in Pavia, in the fight against cancer. A young patient suffering from liver cancer, considered as incurable, was treated using an original technique. Numerous metastases in the liver had been shown by ultrasound and the patient's chances of survival had been estimated at 4 or 5 months. In 2000, the patient had undergone an intestinal ablation as a colon cancer therapy procedure.

In order to try to heal this sick person, an original technology was implemented. Firstly, the patient was treated with a boron phenylalanine solution, an amino acid derivative which has the particular characteristic of binding itself more strongly to cancerous cells than to normal cells, therefore to the tumoral cells in the liver. After a period of incubation, the whole liver was surgically removed, washed, taken to the Pavia nuclear research reactor where it was completely irradiated by neutron bombardment for 11 minutes before being brought back to the hospital to be reimplanted into the patient. 35 minutes elapsed between the two surgical operations. 10 days later, all the metastases were destroyed and the liver tissue slowly renewed itself. The operation's success was only announced the following year, on its anniversary date, and by then the patient, feeling well again, was in the process of recovering a normal liver.

In theory, this technique could be applied to all transplantable organs, the kidneys, pancreas and lungs, and owes part of its success to the absence of any rejection risk. Nevertheless, the liver has the advantage of being one of the few organs that renews itself naturally over time, so facilitating the operation, because a large amount of healthy tissue is subject to destruction.

The properties of Boron 10 have been known for a long time. When subject to neutron radiation, this stable isotope changes into Boron 11, which is unstable and rapidly breaks down into a charged form of stable Lithium 7 and Helium 4 (in fact an alpha particle). Several tests were carried out in the past which involved the treatment by external radiation, using a flow of neutrons, of an organ in which a boron-containing substance was concentrated. Two products containing boron, boron phenylalanine (referred to in the example above), and sodium borocaptate have interesting properties which enable them to concentrate more particularly in tumoral cells. Other products are in the process of being evaluated, but none have been officially authorised, all these tests being currently carried out within the framework of clinical studies. The neutrons used are low energy, but can also act on other atoms which are naturally present in large quantities in cells like hydrogen or nitrogen.

THERAPEUTIC METHODS 101

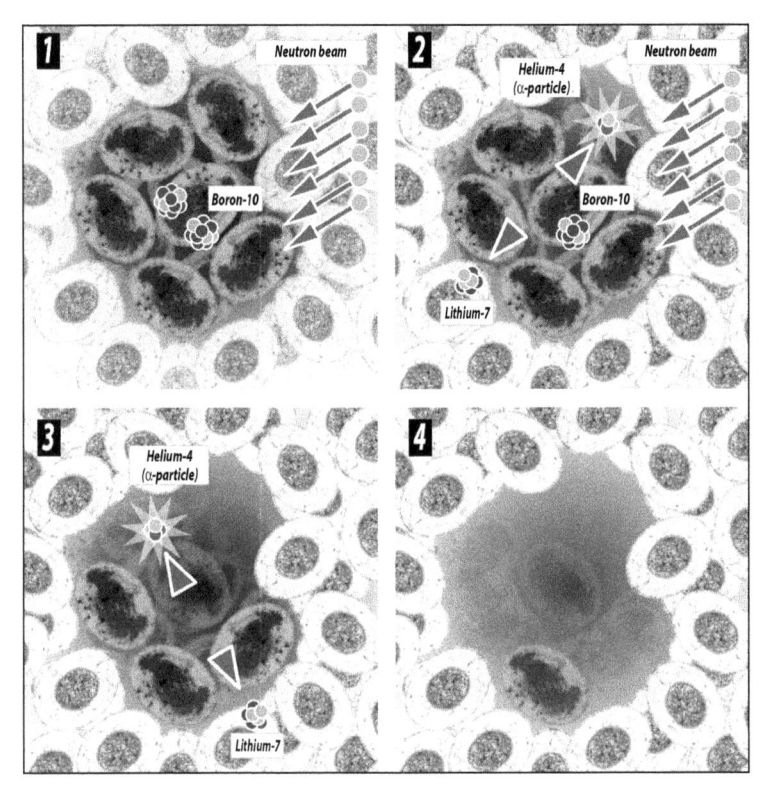

Figure 12. Neutron capture therapy: a (non radioactive) boron-based substance having the particularity of being able to concentrate in tumoral cells is injected to the patient. The tumoral area concentrates in boron atoms and is then submitted to an external neutron beam. When hit by a neutron, the boron generates a neutral atom of lithium as well as an alpha particle that has an immediate cell killing effect.

Nevertheless, the reactivity of these atoms in relation to neutrons is weak compared with that of boron, and any resulting side effects are considered to be negligible, as long as high levels of boron are reached in the tissue to be irradiated.

The technique has been known since the start of the 50's. It is above all of interest in cancers that are complex and difficult to access, such as brain cancers (glioblastoma and astrocytoma). More recently the technique has turned towards the treatment of certain melanomas. It is, however, only effective if a strong concentration of boron-containing molecules is reached in the tumour. As with all radiopharmaceuticals, the important element here lies in the vector's quality and specificity.

However, this technique, known as Boron Neutron Capture Therapy (BNCT), is limited by the lack of available irradiation equipment (a source of neutrons, a reactor or accelerator) and by the side effects of this radiation on healthy organs and cells. The removal of the organ and its irradiation outside the hospital environment gets round these drawbacks. However, the technique remains restricted to patients whose disease is confined to a single organ. If, as well as liver metastases, bone metastases were present, an additional disease-modifying drug would be necessary. The results would be much more uncertain because they would essentially depend on the patient's resistance.

This surgical first opens new vistas for the treatment of certain cancers, and will accelerate the development of new boron-based molecules.

Finally, it should be remembered at this stage that neutron-therapy is originally a technology for the direct irradiation of cancerous tissue by neutrons, that it requires very specific equipment and that neutrons are particularly ionising particles; therefore, great care must be taken in handling them. Neutron-therapy is a radiotherapy technique which, according to our initial definition, is outside the nuclear medicine field.

VII. Radiotherapeutic Substances

The following table gives, in summary, a non-exhaustive list of products labelled with a β^- radionuclide emitter, used in internal radiotherapy and available on the European market. Other yet more specific products, within the scope of very precise indications, are available in the United States. Products with larger markets and intended for a wider group of patients are just beginning to emerge, and are described in more detail in the preceding paragraphs.

Isotope	Chemical form of the radiopharmaceutical	Indications
Erbium 169	Colloidal citrate	Radiosynoviorthesis
Iode 131	Sodium iodide	Thyroid diseases diagnosis and therapy Labelling of molecules for therapy
	Iobenguane(MIBG)	Oncology therapy (pheochromocytomes, neuroblastomes, thyroid medullar carcinomas)
	Esters of saturated acid	Hepatocarcinomas
Phosphorus 32	Sodium phosphate	Palliative treatment of proliferative polycythaemia and thrombocythaemia Treatment of myelocyte chronic lymphocyte leukaemia
Rhenium 186	Etidronate	Palliative treatment of pain linked to bone metastases
	Colloidal sulphide	Radiosynoviorthesis
Samarium 153	Lexidronam	Palliative treatment of pain linked to bone metastases
Strontium 89	Chloride	Palliative treatment of pain linked to bone metastases
Yttrium 90	Colloidal citrate	Radiosynoviorthesis
	Chloride or nitrate	Labelling of molecules for therapy in oncology

VIII. The Dose Issue

Therapeutic radiation is of a different type than that used in diagnosis and the doses administered are often much higher. Is there a dose above which the risk to the patient becomes higher than the benefit gained in the context of therapy? There is no obvious answer.

When a treatment shows a certain level of effectiveness at a given dose, increasing that dose may lead to an improvement in the treatment and to faster healing. For a pharmaceutical product, the feasibility is often almost immediate because in most cases the threshold beyond which the first symptoms of toxicity may appear is tens, if not hundreds of times, the normal therapeutic dose.

In the case of radiopharmaceuticals, given the very low quantity of injected product, the toxicity of the vector plays no role. Therapists find themselves confronted with the problem of radiotoxicity alone, which is well established for a given isotope. The purpose of irradiation is to destroy malignant cells without affecting healthy cells, in the hope that the isotope remains bonded to the vector at least as long as there is an emission of radiation. Therefore, it is the vector itself that should play the main role, allowing a concentration of the radioactivity, and thus limiting toxicity, onto the target cells. Unfortunately this ideal scenario does not exist. The radiopharmaceutical is first of all injected. It circulates in the veins and passes through the liver before reaching its target. Molecules that have not been trapped during this stage re-circulate several times through the blood system, before finally concentrating in the bladder after having been eliminated by the kidneys. This process can last from several minutes to several hours, and during this time the isotope irradiates tissues other than those belonging to the initially targeted tumoral cell. Increasing the dose also raises the risk of destroying part of these surrounding tissues, and in particular, the liver and the kidneys are most affected. The time of residence in these organs must be limited to the minimum.

Clinical studies are in progress to increase the dose by a factor of 10, even 30, in the case of certain radiopharmaceuticals that have already had their effectiveness confirmed in clinical trials at low doses. It is possible in some cases to protect the kidneys by adding substances that act on their function (lysine), but at the price of a new type of toxicity due to these additives. Following initial testing however, it seems that this treatment is partially effective. The studies are carried out with end-of-life patients whose life expectancy is limited to only a few weeks. Depending on the protocols used, a very clear improvement in the patients' state of health has been noted, and in some cases, a remission of several months was even recorded. We are only at the beginning of these studies, and this phase must be considered as a trial and error period which is spurred by the initial successes. On the other hand, the use of high doses creates other logistical problems, but these are of less importance compared with the survival of a patient.

IX. Mechanism of Action – The Bystander Effect

An irradiated cell is affected in different ways, depending on the type of radiation and the energy released. An electron (beta radiation) will have a tendency to create a negative electrical charge, which will move from one molecule to the other until it causes a rupture of the covalent bond, thus generating an unstable radical form. This free radical will be the source of a modification of the DNA chain resulting in the cell's inability to reproduce. Due to the electron's small size, it can move for a great distance, according to its energy, before reaching a target and therefore generating a remote disturbance.

The alpha particle, made up of two neutrons and two protons and several thousand times heavier than an electron, will encounter matter sooner and will have no difficulty in destroying it. In most cases, the particle inserts itself into a bond between two atoms and generates a new, often unstable, entity. If the nucleus is affected, and more specifically its DNA, directly or indirectly, the cell will no longer be able to reproduce and will die.

Several teams of researchers have studied the consequences of bombarding a series of cells with alpha particles while precisely measuring the numbers of particles and cells. A recent example shows that after having irradiated 10% of these cells, it turns out that all of them show a chromosome 11 mutation which is identical to that obtained through direct bombardment. Therefore, irradiating one cell affects all its neighbours. In addition, certain proteins which demonstrate a protective function against malignant changes were produced in greater quantities. Cells communicate with one another, and the death signal emitted by a single cell can be transmitted to other neighbouring cells, leading to their death as well: this is called the bystander effect.

This mechanism has not yet been elucidated, but it offers a definite advantage and partly explains why malignant cells continue to break down even in the absence of radioactivity. The doses of radiation required to destroy a tumour are not necessarily equivalent or proportional to the aforementioned tumour's size; actually, a large part of it may disappear once the destruction mechanism has been triggered, simply due to the proximity effect.

Another hypothesis that may explain the effectiveness of radiation in tumoral cells is based on their breakdown. Ionising radiation triggers a cell destruction mechanism and the formation of fragments recognised as exogenous (foreign) by the organism. These molecular fragments induce the formation of an immune response and the production of antibodies. It is possible that these antibodies may also recognise these same epitopes on the living cells, and thus participate in their elimination. This hypothesis could explain why certain patients show signs of continued healing even a long time after treatment has ended. It could also help understand why others spontaneously healed.

X. The Limitations

Officially, metabolic radiotherapy products are only developed to be administered to adults between the ages of 18 and 65 years. Specific studies are necessary to include elderly people and patients suffering from other equally incapacitating diseases such as diabetes, liver or kidney failure, or patients with severe heart problems. Children are not supposed to be able to benefit from this technology. Finally, this technique will not be used with women who are pregnant or breast-feeding.

In practice, things are quite different. On the one hand, most people suffering from cancer are older than 60 and have parallel health problems. On the other hand, the therapy is often suggested as a last resort after other procedures have failed. Physicians are sufficiently responsible and know better than to take the risk of administering these products to patients at the limit of the therapy scope.

If the side effects of radiotherapy might interfere with those of a parallel, pre-existing illness, the decision is more difficult. The undesirable effects linked to liver, renal or heart failure may prove to be more dangerous for patients than the radiotherapy treatment. Many biological parameters, as well as possible interactions with other drugs used in common treatments, must be taken into consideration. In most cases, death occurs as a consequence of a general organism weakness, itself induced by one of these treatments.

So far, we have limited our descriptions to the development and use of radioisotopes in relation to the patient. Nevertheless, patients will only be in contact with radioactivity once or twice during the course of the treatment. Medical staff on the other hand, are in permanent contact with several patients undergoing treatment, and are therefore subject to this radiation on a daily basis. Protection equipment is of course implemented in each hospital. Nurses and physicians are also equipped with dosimeters and are monitored regularly, as with any activity linked to radioactive material. The rules of protection described elsewhere obviously apply to medical staff as well. They can be summarised in three words: time, distance and protection. Reducing the time spent in a radioactive environment as much as possible, therefore also the time spent with patients being treated, increasing to the maximum the distance separating them from the radioactive emitter, therefore from the patients, and finally taking maximum advantage of the means of protection provided, such as partitioning, windows and lead-lined equipment.

Very strict rules are put in place for the release of a patient that has been treated, and have more to do with their waste products' radioactivity level (urine, faeces, perspiration) than with their own body radioactivity. In all cases, the risk to the family environment will always be less than that run by medical staff. Some countries like the United States prefer to minimise the irradiation risk to their staff by releasing patients sooner, considering that contamination linked to waste products is of lesser importance. The two concepts are debatable.

These considerations put the emphasis on other limiting criteria, such as the access to care through the number of equipped rooms, and the availability of competent staff, whether they be doctors or nurses. This technology will not be able to develop unless the infrastructure necessary for its setting up meets the demand.

Summary

A radioactive substance that targets a specific tissue will provide a usable image of this area if the radionuclide used is a gamma or beta-plus emitter, or will partially destroy this area if it is a beta-minus or alpha emitter. In theory, any substance capable of giving a specific image of a tumour can be converted into a therapy product if the diagnosis radionuclide is replaced by a therapy radionuclide.

In practice, things are not that simple, because the ideal therapy radionuclide – the one that will only destroy the cancerous cells – does not exist and account must therefore be taken of all its physical and chemical properties.

Iodine 131 has become the reference treatment for thyroid cancers due to its specificity to thyroid cells, whether cancerous or not.

More complex mechanisms help explain the effectiveness of Strontium 89, Rhenium 186 and Samarium 153 in the **palliative treatment** of pain from bone metastases. Injected locally, Erbium 169, Rhenium 186 and Yttrium 90 are used in the treatment of rheumatoid arthritis. This method is called **radio-synoviorthesis**.

Radioimmunotherapy aims to destroy cancerous cells produced as a result of an immune system dysfunction, targeting more particularly the antigens of abnormal cells.

In theory, most receptors located on malignant cells can be targeted by vectors onto which a radionuclide has been grafted. And in practice, several tens of these new labelled molecules have been successfully tested with different types of cancer.

Alpha emitters present an even greater therapeutic potential, but their use still requires a number of developments. A particular technique, the therapy by **neutron capture**, could open up new perspectives, but will probably be limited due to its complexity.

The mechanisms explaining the effectiveness of radiotherapeutic substances are not as simple as it would first seem. Apparently cell repair continues to operate well after the radioactivity has disappeared, and not necessarily in proportion to the dose administered (bystander effect, dose effect).

CHAPTER VII

The Development of Radiopharmaceuticals

Metabolic or targeted radiotherapy applied to oncology has led to some very convincing and promising results. Yet, the successful initial application of iodine in the treatment of thyroid cancer was followed by a long, and apparently unfruitful, period at the therapeutic level. However, these years enabled the mechanisms involved in the action of radioisotopes to be better understood. They also helped evaluate which isotopes were best suited to therapy, improve imaging techniques, develop labelling methods and, above all, find the appropriate means to produce the complex molecules that could serve as vectors. The tools and the technology are now available to researchers. Once again, they must be allowed time to develop ideal molecules. The first effective second generation molecules only really appeared at the start of the twenty first century. A number of radiopharmaceutical products are in the course of evaluation and other avenues are being explored in research laboratories. Nevertheless, the development of a drug requires a great deal of time and money. The radiopharmaceutical aspect brings an additional constraint, which can however present original and particular advantages in certain circumstances.

The following paragraphs relate the different stages, in sequence, of a drug's development, and more particularly a radiopharmaceutical product, starting from the chemists' discovery of the very first molecular entity to the authorisation of the product's marketing, including the verification stages of its biological, pharmacological and toxicological properties, animal testing, then human testing, and finally the preparation of the final administrative dossier.

Which Market?

Depending on the therapeutic indications, a new molecule arriving on the market will have cost between 200 and 900 million euros and will have taken from 9 to 12 years of work. In order to be profitable, a new product must generate, in average, a turnover of the same amount within each year following its marketing. Taking the financial stakes into account, it is mandatory to choose the right field before researchers commit themselves.

Each company has a strategic marketing unit which evaluates the most promising fields, by looking into the future in order to estimate real medical needs. Companies carry out market research to assess the potential development of the future drug. Requirements in the pharmaceutical field are also linked to the sudden appearance of new pathologies (aids, hepatitis), to behavioural changes (excessive tobacco and alcohol intake, leading to an increase in the number of cancers) and to bad eating habits (leading to obesity). An increase in certain risks (a higher number of diabetics), or the ageing of the population (Alzheimer, Parkinson's diseases) also play a part. In addition, these markets depend on changes in legislation, regulations and policies (care and reimbursement level for the drug). The value of the market is significant because the cost of developing a product will be practically the same whatever the chosen indication may be. Therefore, pharmaceutical companies fight over profitable niches such as neurology and cardiology, while at the same time neglecting diseases that affect a large number of people (malaria, tuberculosis) but which occur in countries with little or no financial capacity. The pharmaceutical industry has no philanthropic vocation and is obliged to perform financially in the same way as any other profit centres (such as food or automobile industries).

Even if the niche is profitable, the marketing strategy must be capable of forecasting the market share that will be taken by the company in the field in question, without being aware of the progress achieved at the same time by the competition. The pharmaceutical industry is without doubt in the most competitive market there is, with the added constraint of a very imprecise economic forecasting. The bet is always very risky. Nevertheless, it is on this basis that choices are made and objectives are set for researchers.

The decision to develop a new radiopharmaceutical product must in principle follow the same rules. Nevertheless, potential markets and production constraints limit profits to rarely more than a tenth of the figures mentioned above. Consequently, these products can only be profitable if development costs, or else competitive risks, are reduced proportionally. Certain specific parameters show that these developments remain possible still.

I. The Molecule Discovery Phase

A researcher in charge of "developing a new drug for a defined treatment" quite simply starts with copying: copying nature, and copying the competition. Then, he transforms with a view to improve, of course, and also to meet the pharmacologists wishes.

Chemists alter molecular forms and structures on the basis of their scientific knowledge and experience, but also their intuition. Each new molecule is tested by biologists. As results progress, chemists try to improve the product's properties in order to obtain a family of new compounds with a certain level of effectiveness. The new molecule set must be able to be patented, therefore it must be original.

This first stage of a vector's development is in fact independent of a radiopharmaceutical's: there are so many molecules available on the market today that wanting to develop one's own vectors is pointless. At the very most, the chemist will be able to modify the structure of the molecule in order to improve a particular property. On the other hand, he will need to find a means of incorporating or grafting a radioisotope onto it without affecting its biological properties.

Recent estimates show that, on average, only one molecule becomes a drug out of 4,000 to 5,000, and synthesising as many requires the lifetimes of several chemists. As radiochemists only use these marketed molecules as a starting point, their chance of transforming them into radiopharmaceuticals is that much greater, and therefore the time period for the "discovery" of a radiopharmaceutical is reduced by just as much.

II. Pharmacological and Preclinical Studies

While chemists are kept busy with their reactors, pharmacologists strive to develop a test *in vitro* which will enable them to confirm whether the new molecules are more or less active as compared with the natural substance, or the competitor's molecule, of reference. A traditional molecule will go through a series of quick and not very stringent tests, allowing the best to be selected (screening). These tests are essentially carried out using living cell extracts such as receptors or enzymes. The ability of a molecule to attach itself to a cell (to bind) or to block or activate an enzyme will also be measured.

A slightly more sophisticated, more extensive, and therefore more costly test eliminates yet more molecules among the tens of interesting ones. The information obtained at this stage simply confirms that the molecules from this initial series interact in the mechanism that enables the pathology to be described, but it doesn't specify whether that interaction helps healing or, on the contrary, makes the disease worse. A partial answer to this question can be obtained by testing the molecules' activity on an isolated organ, like a heart for a future cardiology product, or part of an intestine for a drug being developed for gastroenterology or immunology purposes. At this stage, the biochemist's imagination comes into play. Products intended for neurology, for example, are not necessarily developed using brain extracts.

The third and last stage involves a few dozens of molecules. It is the most thorough because it makes use of preconditioned animals, simulating the indication being targeted.

In oncology, the most usual method consists of treating mice which carry tumour transplants of human origin. These stages are extremely costly, both in terms of money and time.

During the development of radiopharmaceuticals, a large number of these steps can be reduced. The reason is quite simple: the radioisotope grafted onto the molecule plays its role as a tracer, and imaging allows the molecule to be located immediately. The best substances are selected on the basis of those that show the greatest aptitude to concentrate on the specific tumour, or on any other organ or tissue being targeted.

III. Pharmacokinetics

Pharmacokinetics studies define the time during which the drug is going to reside in the cells, and more generally in the organism, and metabolic studies identify the transformation processes, and also the elimination of the molecule.

Absorption, distribution, metabolism and elimination are four terms that define the drug's evolution once swallowed or injected. Studying these parameters provides information on the speed at which the product is distributed in, then eliminated from, the body (pharmacokinetics), and also on the mechanisms of this distribution (pharmacodynamics). These values help determine whether the product should preferably be given orally or intravenously. Depending on these conclusions, it is the galenist's responsibility to find a suitable formulation. These formulation specialists must take the target in which the drug needs to concentrate, and the speed at which it must dissolve, into consideration. A whole range of media enable them to accelerate or delay the dissolution time and decide whether the capsule should disintegrate in the stomach or in the intestines. In addition, they indicate when a patch is more suitable than an injection, while confirming that these combinations do not degrade the quality of the product over the course of time. The formulation of a drug goes back and forth an incredible number of times between the galenist and the pharmacokinetic scientist, in order to optimise both the distribution of the medication in the body and the end product's longevity on the pharmacist's shelves.

In the case of radiopharmaceutical products, following the distribution of a labelled molecule in an animal is just as easy as following its evolution after ingestion or injection. As far as the formulation is concerned, radiopharmaceuticals are essentially administered in the form of intravenous injections, using a syringe, and more rarely orally (iodine for thyroid cancer therapy).

IV. Toxicological Analysis

A great deal of information concerning the toxicity of products can be deduced from their pharmacological and biological properties, as well as on the basis of their structural analogies. On the other hand, theory is insufficient, and the risk too great, to dare expose an individual to a new product without additional testing. A new medicinal substance is never administered to a human being before the molecule's toxicological profile is known. Unfortunately, there is currently no other alternative as reliable as animals. Certain criteria can be predicted but remain insufficient. Three levels of testing are necessary. Acute toxicity studies examine the effects on the animal up to the point where the maximum tolerable dose is reached. Long-term effects can be observed with chronic toxicity studies, where an average dose, higher than the normal therapeutic one but not fatal, is administered over several weeks. Finally, reproductive toxicity testing is necessary in order to demonstrate that the substance does not lead to any malformation or genetic modification in progeny, even to the second generation.

As it is quite difficult to directly transpose animal testing results onto man, this information on toxicity is only informative in character. Basic precautionary rules and regulatory authorities' recommendations oblige the pharmaceutical industry to carry out these tests on two different animal species.

The toxicity of a radiopharmaceutical is characterised by two aspects. Firstly, all the radiological risks linked to the absorption of a radioisotope must be taken into consideration. This subject has already been discussed in detail in the chapter concerning radioactivity and radiation, and this toxicity is well-known and well-controlled. Once an isotope's distribution in the organism is identified, it is possible to evaluate its radiological impact on the organ in which it is retained and so deduce the maximum dose that can be injected, and which should not be exceeded.

It is perfectly obvious that this injected dose must take the radioisotope's specific activity into account, that is to say the relationship between the activity of this radionuclide and the total mass of the element present. A radioisotope is frequently accompanied by its equivalent stable element. In a way, the radioactive molecule is diluted in an environment containing this same molecule in its stable form. The term "carrier" is used to define this stable fraction.

The second aspect relates to the cold part of the molecule, which becomes insignificant however when compared with the risks associated with the radioactive part of the molecule. In most cases, the toxicological profile of the vector is known, since the molecule is often the product of traditional pharmaceutical research which itself provided all the toxicological data.

The quantities of biologically or radiologically active material injected into a patient are extremely low. For example, the quantity of iodine injected in order to carry out thyroid scintigraphy corresponds to a thousandth of the iodine dose that we absorb daily with our food. The toxicological effects associated with this extra iodine are invisible. This example may be applied to all injected products. Whichever the vector used may be, it is always at a considerably lower concentration level than the point at which the first signs of toxicity are likely to appear. This affirmation is true to such an extent that it is possible to envisage labelling toxic substances with a radioisotope in order to inject them into man, knowing that the toxicological risk remains fully under control. The side effects linked to the radioisotope grafted onto the molecule will always be greater than the intrinsic toxicity of the vector, at the doses used.

All that remains to be confirmed is the absence of abnormal immune reactions (allergies), and, in the case of antibodies, the formation of human antimouse antibodies (HAMA), human antichimerical antibodies (HACA) or human antihuman antibodies (HAHA).

With these findings, the development of new nuclear medicine products can be envisaged on a completely different level: a number of molecules studied in the context of developing new pharmaceutical products display remarkable biological profiles (very high level of binding), but unfortunately have to be rejected due to their high toxicity. Due to their specificity, they make excellent vectors for future radiopharmaceutical molecules, as their toxicity level becomes increasingly insignificant. The major pharmaceutical companies' research centres count many abandoned molecules in their drawers which could serve as a basis for new radiotherapy products' families.

Insofar as a radio-labelled molecule can be seen directly through the organs, its pharmacological evaluation is also accelerated, because all the information is visible via images of a complete animal. Tests are carried out more quickly, they are less numerous and the data obtained can be used immediately. In the case of a traditional pharmaceutical substance, it takes two years on average to obtain the necessary data before moving on to man. This same information is available within barely six to nine months with radio-pharmaceuticals.

V. Phase I Clinical Studies

Clinical studies cover all the tests carried out in humans. With the help of several healthy volunteers, phase I studies attempt to show that this new product is harmless.

On the basis of the data collected during the above-described preclinical studies, whether or not to carry out human testing is one of the most important decisions that need to be made. This is for three reasons. Firstly, injecting a product into a human being for the first time is not a trifling matter, and the whole company's responsibility is at stake. This decision is taken with the participation and support of toxicologists and medical experts specialising in this field. Secondly, doubts concerning the toxicology results or the effectiveness of the molecule may still exist, and it is sometimes preferable to confirm these results by supplementary preclinical studies. Finally, committing oneself to the clinical phase, involving humans for the first time, is the most costly development phase. The right molecule must be found, and the right target must be hit first time.

Before even finding out if the new molecule has a significant therapeutic advantage, both the fact that it is harmless to humans and an idea of the maximum dose that can be tolerated need to be confirmed. At this stage, only data obtained from animals is available for reference. This first study must be carried out on healthy male volunteers. The main objective is to detect any undesirable or even harmful effects of the molecule in humans, which were not detected in the animal. In general, doses a hundred times lower than those that caused side effects in animals become therapeutic doses. The clinician starts by giving even lower doses to the volunteer, increasing them progressively. Regular listening to and constant monitoring of these volunteers help establish the dose at which these first symptoms systematically appear, information that the animal could not communicate (pain, dizziness, migraine, inadequate taste, etc.) In general, this dose also corresponds to the maximum dose which should not be exceeded. In order to limit other side effects as much as possible, these first tests do not involve women, children, elderly people or people who are too frail. Paradoxically, drugs intended for women only, such as the contraceptive pill, are also tested on men in the first instance. The formulation used (gel, capsule, pill, patch, injection or suppository) will remain the same for the entire study, even if a few more minor changes are still necessary at this stage.

Healthy volunteers are selected by a team of specialist clinicians, on the basis of very strict criteria, including their non-participation in other similar studies in order to avoid an unrecognised interaction with other molecules in the course of development. Tests are carried out in a hospital that is fully equipped for intensive care. Volunteers in phase I clinical trials are remunerated for their active participation in these experiments. Their status is protected by law and by the Helsinki agreement.

In general, the results obtained on the basis of a dozen participants are sufficient to decide whether to move on to the next phase. Along with collecting subjective data from the volunteers themselves, a number of parameters are recorded (electrocardiograms, electroencephalograms, etc.) and analysed (blood, urine, faeces, saliva, sweat, etc.) Two or three optimal doses are deduced from these studies, as well as the maximum dose which should not be exceeded. Finally, the type of formulation used is confirmed.

VI. Phase II Clinical Studies

As they move on to clinical phase II, clinicians aim to demonstrate that the product has a positive effect on the pathology, and seek to estimate the ideal dose that should demonstrate the new drug's effectiveness.

Whereas recruiting a dozen of volunteers through the press for a phase I clinical study is easy enough, it is more difficult to start an experiment with people who are sick. Phase ll consists of confirming the effectiveness of the new drug on a small number of patients, generally about forty, who have agreed to take part in the experiment. The study is carried out in collaboration with physicians on the basis of a very strict protocol developed jointly. The first study, which generally takes place in hospital, takes into consideration most of the reference parameters used in phase I (heart, neurological, biological), but its main objective is to analyse the evolution of the illness.

Experience shows that, depending on the environment, in some cases a clear improvement in the patient's condition is noted, even in the absence of treatment. This is called the placebo effect. When patients feel cared for, have confidence in the team providing care, and believe in the effectiveness of the treatment, their physical condition may improve to such a point that the drug may seem to be of no use. Conversely, failing morale, a depressed psychological condition, misunderstanding the intentions of the people with whom patients communicate and a pessimistic environment may aggravate a situation, even when the best of treatments is being applied. This is the nocebo effect. These effects may be very significant. They account for more than 40% in the case of treatment of all gastroenterological problems, and in particular ulcers, and more than 80% in the case of neuropsychological problems. In other words, this means that an antidepressant drug, for example, will only have a therapeutic effect in the remaining 20% of patients and that the administration of sugar, combined with the therapist's conviction, could practically improve the condition of 80% of the other patients. As it is impossible to test all these psychological parameters, physicians responsible for clinical studies have had to incorporate this factor into their protocols.

Each study is carried out using a control group. This reference group is selected following the same criteria as for the group to be treated, and it receives an inactive form of the drug which is in all other points identical to the medication being tested, called the placebo. Neither the patient, nor the doctors, nor the care staff know who will or will not receive this drug. Selection is random and the names of the patients who have really been treated will not be revealed until the very end of the study, when the results are collated and analysed. This is a double blind study which is the only way to bypass the environmental influence on the treatment.

It should be noted however, that studies using a placebo do not apply to serious pathologies, as stopping an effective treatment would not be ethical. It is obvious that physicians do not allow the life of a patient to be put in danger. The traditional treatment is maintained for the two groups of patients, and the new treatment is given to one group only as a supplementary course, to see whether it brings additional effectiveness.

In parallel to the verification of the effectiveness of the treatment, physicians try to determine the ideal dose. In order to do this, they may need to separate the group being treated into sub-groups to which different doses are given. If results show that another dose could have been employed and would have led to better results, or that a change in formulation could have been more effective (gel capsule instead of pill), the study is started again. Phase II results determine whether the last clinical phase can be undertaken.

The length of the study depends on the parameters being followed. As it is about effectiveness, physicians will above all seek to find out if the product acts quickly and effectively on the illness. In some cases, results are obtained in a few days. In oncology, the main objective is often linked to the patients' survival. They are monitored, therefore, for a long period after the treatment has ended, generally equal to the average life expectancy for this category of patient.

Diagnosis radiopharmaceuticals come into the first category with very fast results. The image acquisition is immediate, and can be interpreted within a few hours. Physicians only have to wait, at the very most, for additional parameters, biological analysis for example, in order to confirm the test's validity. In the case of diagnosis, the placebo effect is rarely mentioned. On the other hand, the protocol is written in such a way that the physician in charge of the evaluation can only read the images in a blind form, without knowing the patient or their history.

Therapy radiopharmaceuticals come into the category of products requiring a long period of analysis before their effectiveness is confirmed. Double blind treatment is particularly difficult, and little used, because the standard reference treatment rarely consists of another radioactive treatment. It is quite easy for patients to identify when they are being injected with a radioactive substance, if only because of the special environment in which the test is performed. On the other hand, metabolic radiotherapy products are only presently administered to patients for whom all other traditional methods of treatment have unfortunatly failed. This complicates these studies' organising, and the clinicians who write the protocols have to bear this in mind. Nevertheless, taking into account the fact that it is possible to verify the product's distribution by imaging techniques, results are obtained more rapidly. Their analysis is also accelerated and, due to the precision of the method, the overall number of patients to be treated can be reduced.

As injecting radioactive products into healthy volunteers is neither authorised nor recommended for ethical reasons, phase I and II clinical studies are combined together. Side effects studies are directly carried out on patients, and this reduces the number of patients involved in these phases accordingly. Clinical phase III is consequently sooner carried out.

VII. Phase III Clinical Studies

Phase III studies demonstrate the actual effectiveness of the product in a large number of patients, along with its superiority as compared to the current reference treatment and the absence of any widespread side effects.

This phase is a key stage in any development because products are very likely to be marketed one day. One molecule in five or six passes the phase I stage. This proportion is the same for phase II. A radiopharmaceutical has one chance in two or three of passing the first clinical stages. Phase III serves to confirm both the declared effectiveness and the absence of side effects in a much greater population. Depending on the indications, a clinical phase III study requires from 500 to over 4,000 patients. Two distinctly parallel studies need to be carried out, which are called pivotal studies. If the product's effectiveness really is demonstrated in the course of phase II, a placebo group is no longer necessary. On the other hand, in order to be marketable at a later stage, the product must show a clear advantage over the commercially available reference drug at the time of the study. Therefore, clinical studies are constructed in such a way that the results of a group receiving the new drug can be compared with those of a reference group being treated with the traditional products (preferably the best on the market, called "gold standard"). In order to avoid any external influence, this study is also carried out as a double blind. Formulation scientists are sometimes obliged to repackage commercially available drugs to make them look like the new treatment under evaluation.

The final product used in this study must be in the same medicinal form as when it eventually gets marketed. Chemists and formulation scientists take advantage of the phase I and II period in order to develop a method of synthesis, a final formulation and a method of production that can no longer be changed, except by carrying out the phase III study again.

Taking into consideration the number of patients involved in this phase, it is also the most costly. About one third of the overall cost of developing a drug is absorbed by this clinical trial phase.

In parallel, to anticipate certain questions which are bound to be soon asked by the authorities, and to ensure the safety of the product for a wider public, some additional phase I level clinical studies will be carried out. The term phase I does not always correspond to the first stage, nor is it automatically linked to the involvement of volunteers, but it defines the number of parameters to be monitored in an extremely rigorous fashion. Depending on the circumstances, the drug's effect on certain sub-groups of patients is studied, such as individuals suffering from liver or renal deficiency, or patients of a particular type, for example suffering from obesity or diabetes. Sometimes the harmless nature of the drug in relation to elderly patients needs to be confirmed and, if the product is to be given to children, a study needs to be carried out in a paediatric environment.

Another important mandatory complement consists of including, within the group of patients participating in a phase III clinical study, a proportion of individuals corresponding to the ethnic minorities of a particular region. It may happen that certain treatments are less effective in people of a different ethnic origin. In order to demonstrate the contrary, a study carried out in France should include about 10% North Africans, and in the United States, the same proportion of black people. The Japanese Government requests that a full phase III study be carried out, preferably in Japan with an Asiatic population.

American regulations oblige the industry to demonstrate the product's effectiveness in two distinct phase III studies, commonly known as the pivotal studies. For these two studies to be of benefit at the marketing stage, they slightly differ in terms of patient selection, expected results criteria, and reference products. Patient distribution is revealed at the end of the phase III study. This crucial stage in a product's life, called the unblinding, gives confirmation of the sub-group in which the drug proved to be the most effective. It is only at this time that it is possible to find out whether the product is effective or not in comparison with the control group. This stage virtually signs either the product's marketability, or its death warrant. At this stage the company will have already spent several hundred million euros.

Due to the particular nature of radiopharmaceuticals, the number of patients included in a phase III study is relatively small. This parameter does not affect the time involved, but it has considerable impact on the cost. In consequence, a company developing radiopharmaceuticals can develop a product aimed at a much smaller population, and thus target illnesses that are a little less common.

Data Protection

A new drug is only of value to a company if the latter can keep a monopoly over its product and exploit it for as long as possible. The only effective protection consists of holding a patent. Any new molecule which displays original pharmacological properties can be protected for 20 years, to the inventor's advantage. During these 20 years, no other company can exploit the invention in the country where the patent was filed without authorisation from the patent owner. 20 years seem both long and short. Indeed, once the 10 years required for the product's development are deduced, the research's results must be exploited and bear fruit within 10 years only. For, after this period, the product will enter the public domain to become a generic product, and any company that is sufficiently well-equipped will be allowed to market it without having to expend any of the investment that was initially required for its discovery. Patents are applied for as late as possible, because each year gained concerning the patent date is an additional year of sales without competition. On the other hand, waiting too long is very risky, because if the niche is profitable the molecule may in the meantime be discovered by a rival team which could then apply for a patent itself. The actual process of applying for a patent gives an additional year of confidentiality to inventors, to allow them to improve their discoveries. The patent is then assigned for publication, approximately 18 months after the initial application was made. A personal patent may be invalidated if the same molecule appears in a competing patent which was applied for prior to this period. During this time, research efforts will have been undertaken for nothing.

VIII. Regulatory Issues and Registration

If the results are conclusive, the company confirms its interest in marketing the product. Between 8 and 10 years will have elapsed since the day when it was decided to start this new project. A marketing authorisation application is made to the competent health authorities. This very full dossier covers all the information linked to the new product, from its manufacture to all the observations collected individually from each patient or volunteer while it was being developed. The final document represents tens of thousands of pages that need to be collected, checked, analysed and summarised.

Obviously, this laborious task must be carried out at the same time as the work's progress. But the results synthesis can only be done once the dossier of the last treated patient has been analysed.

After pagination, the dossier can be photocopied and the hundreds of kilos of paper are sent to the relevant authority. For several years now, an electronic copy (CD-Rom) must also be supplied to the authorities for quicker data processing and analysis.

As far as possible, a pharmaceutical dossier of this type is filed with the three major regions corresponding to the three major world markets. For Europe, a single dossier must be provided and filed with the EMEA (European Medicines Evaluation Agency) in London. The FDA (Food and Drug Administration) decides on American dossiers (United States), and the Koseisho deals with Japanese ones. Given the complexity of the work, it is preferable to file the dossier with one of the major regions (Europe or the United States) before filing it with a second region in order to be able to provide a corrected dossier and limit the questions, once exchanges with the first agency have taken place. Dossiers should only be filed in Japan in third place, due to this country's protectionist policy which leads to additional and extremely severe constraints.

The agencies must be warned several months in advance of the arrival of a new dossier, so that they can organise themselves and start working on it as soon as the documents are handed over along with payment of the evaluation costs (232,000 euros for the EMEA in 2005).

The authorities undertake to analyse the dossier within a given deadline. Payment of the charges equates to a contract, with very precise due dates concerning the work they have to provide. Initial comments and questions must be returned on a set date. In certain extreme circumstances, an additional clinical study may be requested. In this case the project may be postponed for several years.

In practice, an average of 18 months should be allowed between filing the dossier and obtaining the marketing authorisation. After this authorisation is obtained, due to the workings of recognition between countries, other applications will be filed with countries which are not covered by the three major markets. Up to 3 years can easily pass before a drug first marketed in the United States is authorised in Europe, and vice-versa. It can even take up to 10 years for it to become available in Asia or South America. At this stage, procedures are the same for pharmaceutical and radiopharmaceutical products.

IX. Marketing

Obtaining an authorisation is the most important stage before marketing. Nevertheless, several administrative barriers still have to be overcome. In the case of a product intended for hospital use, negotiating its price and obtaining its reimbursement rate require local authority approval, country by country. These procedures take a while, and it will be a few more months still before the medicine can be prescribed and patients can benefit from this new treatment.

As radiopharmaceutical products are only sold in a hospital or clinic environment, price discussions are very limited. Then, even if a good sale price has been negotiated, the sales volume is directly linked to the budget allocated to hospitals by the local government, thus limiting the industrial income.

More than 10 years separate the initial decision to break into a new niche from the moment the drug can be sold on the market. Each time reduction in any one of the development stages has obvious economic repercussions. Due to the simplification of some of the stages, and to the reduced number of patients involved in the development of radiopharmaceuticals, such products can be marketed within 7 to 9 years of the studies being started. The budget swallowed up in this development will remain limited between 40 and 100 million euros for an average potential annual turnover at the level of 50 million euros.

X. Post-marketing Authorisation and Drug Monitoring

As with all drugs, radiopharmaceuticals must be monitored to detect any side effects that may only appear once this type of treatment has been administered to a greater number of patients (pharmaco-vigilance). In fact, expected side effects are extremely rare, and of lesser importance, especially when compared with the illness they are supposed to follow up or treat. All this information is collected by the authorities and highlights any possible undesirable or harmful effects of a drug which could not be assessed during clinical studies. Some side effects only appear in a ratio of one in a million, and therefore very often remain invisible during clinical phase lll, which only included several thousand patients. If the drug is distributed to millions of sick people, dozens of people may be affected by these previously unobserved side effects. It is important to collect and centralise this information. Depending on the seriousness of the effects observed, the drug may even be withdrawn from the market.

A drug may be sold when it has shown itself to be effective in a particular sub-group of patients. It is then marketed with this single indication. Physicians soon start to ask for an extension of this indication to another sub-group of the population (another age group for example). In order to be able to sell the product to a wider public, its effectiveness and safety must be demonstrated through an additional study. The purposes of so-called phase IV clinical trials, which are only carried out once the drug has been made available on the market, are to explore the dosages suitable for selected subpopulations, to study the different administration regimes or to check compatibility with other treatments available on the market (and administered at the same time). Undeniably, product prescribing is safer with these studies.

Extending the indication itself, in other words using the same medication to treat an illness for which the drug has not been tested, requires a new demonstration via a phase III clinical study. In consequence, a new dossier must be filed.

Summary

Before it is marketed, a new drug undergoes a development phase consisting of a series of tests, first on animals, then on humans. These tests require ten years of work and an investment of several hundred million euros.

All molecules roughly follow the same development sequence:

– **Chemical synthesis:** after having developed the molecule and its synthesis, chemists have it tested on simple *in vitro* models (in test tubes) in order to check its biological potential for the mechanism of the indication being considered.

– **Pharmacology:** this second stage consists of verifying if complete and isolated organs provide a positive answer to this molecule.

– **Preclinical studies:** the substance is tested on animal models in order to check the positive effect of the substance on a particular pathology.

– **Toxicology:** at this stage, it is important to determine the toxicological risks linked to the substances, both at very strong doses, and at weak doses but administered over a long period.

– **Galenic:** the method of administration (formulation) of the future drug is developed.

– **Clinical phase I:** if all the initial tests are positive, this substance can be injected into healthy volunteers, in order to check for the absence of side effects.

– **Clinical phase II:** this consists of checking the product's potential in the first patients, but above all of finding the ideal dose that should be used for the treatment to be effective.

– **Clinical phase III:** these trials must demonstrate, in a very large number of patients, that the drug is effective at the dose recommended in the previous clinical trial.

– **Registration:** before being able to market the product, the necessary authorisations must be obtained once the quality of the studies and the validity of the results have been verified by competent experts.

Radiopharmaceuticals are subject to these same development constraints and steps. As a consequence of the imaging technology, some results are obtained immediately (the image itself), and because the quantities injected are extremely low as compared with traditional drugs, the whole toxicology study phase can be reduced. Therefore, the overall development period is shortened by a year or two.

Once marketed, the drugs continue to be monitored very strictly (pharmaco-vigilance), and radiopharmaceuticals are no exception to this rule.

CHAPTER VIII

The Production of Radiopharmaceuticals

Radiopharmaceutical product status is obtained on the basis of very well defined production quality criteria: final product quality that is beyond reproach and which corresponds to the specification described in the dossier filed with, and approved by, the various drug agencies. Also, the manufacturing process must be reproducible from one batch to the next.

The radionuclides and labelled substances which are produced using a method other than the one that is officially filed with the authorities are called "radiochemicals". They cannot carry the label "radiopharmaceuticals". When a new product is being developed, it is obvious that the substances used also have the status of radiochemical product. Nevertheless, those that are injected into humans in the context of a clinical study are subject to the same type of controls as radiopharmaceuticals. Therefore, they must follow the same criteria guaranteeing product quality to prevent any risk of harming or endangering patient lives.

I. Definitions

Due to radioisotope decay, the composition of a radiopharmaceutical product changes over time. It is therefore necessary to define several measurable parameters, and in particular those informing physicians on the precise quantities of radioactive substances actually injected into patients.

The **specific activity** of the radioisotope corresponds to the ratio between the activity of this radionuclide and the total mass of the element or molecule present. A radioisotope solution is called **carrier free** when only radioactive atoms are present, even in an extremely dilute solution. Specific activity is expressed in becquerels per mass unit. This concept of radioisotopic concentration is essential for labelling vectors. Coupling an isotope with a vector results in a reaction which can only be effective and return a very good yield in the absence of competition between hot and cold isotopes.

Specific activity should not be confused with **radioactive concentration** (or volumetric concentration). The latter determines the degree of radioactive substance per volume unit (expressed in becquerels per volume unit).

A radiopharmaceutical must be produced in a precise time frame. Taking decay into account, the radioactive concentration is always higher at the end of production than at the time of injection. In order to take transport time into consideration, a radiopharmaceutical is delivered with a certificate indicating the radioactive concentration at a precisely defined time (usually close to the time of administration). One then speaks of the **calibration date** and of activity or dose on calibration. On the basis of this value, physicians can precisely calculate the volume fraction to be taken when the patient is being injected, or at the radio-labelling stage. The calibration date should not be confused with the **expiry date** or shelf-life, important criteria for any pharmaceutical product. Decay is accompanied both by a dilution effect and the appearance of new impurities, including radiolysis products and decay-produced isotopes. Stability studies carried out on samples have defined a specific time period, at the end of which the product no longer meets its own quality criteria. These parameters must take storage and transport conditions into consideration, particularly storage temperature and any possible interaction with the container and other materials contained in the flask. The expiry date may be extended artificially by adding stabilisers (excipients), but excipients won't actually have any influence on the radioisotope's decay.

II. The Principles behind Radionuclides Production

Radionuclides can be produced in several ways, using a generator, a reactor, a cyclotron or a linear accelerator. Extracting products after having obtained them in a thermal reactor, or separating them from fission products are other ways to produce radionuclides that can be used in nuclear medicine.

Two major difficulties limit the methods of production. Firstly, handling radioactive substances requires a suitably safe environment that is therefore very costly. Secondly, nuclear medicine can only use radionuclides of very high purity, requiring optimised methods of separation and purification. Generally speaking, most radionuclides to be used pharmaceutically are easily accessible. In fact, products used in nuclear medicine have only been developed on the basis of radioisotopes that are easily accessible. Nevertheless, by virtue of their short half-life, they require rapid transformation. The production and purification parts of the process are obviously included in the period of use of the radioactive material, as because of its rapid decay it constantly generates daughter isotopes which must also be considered as impurities.

Particle Accelerators

A radioactive substance can be formed by bombarding a stable substance with charged particles in a linear or circular type accelerator. The tools most frequently used are circular accelerators, also called cyclotrons. But whatever the technology used, the same type of radionuclide is formed by the bombardment of an identical target with the same type of charged particles.

Thus, in a cyclotron, charged particles of low mass, such as protons, are accelerated in a circular trajectory until they reach high energy. These particles are used to bombard a specific target which

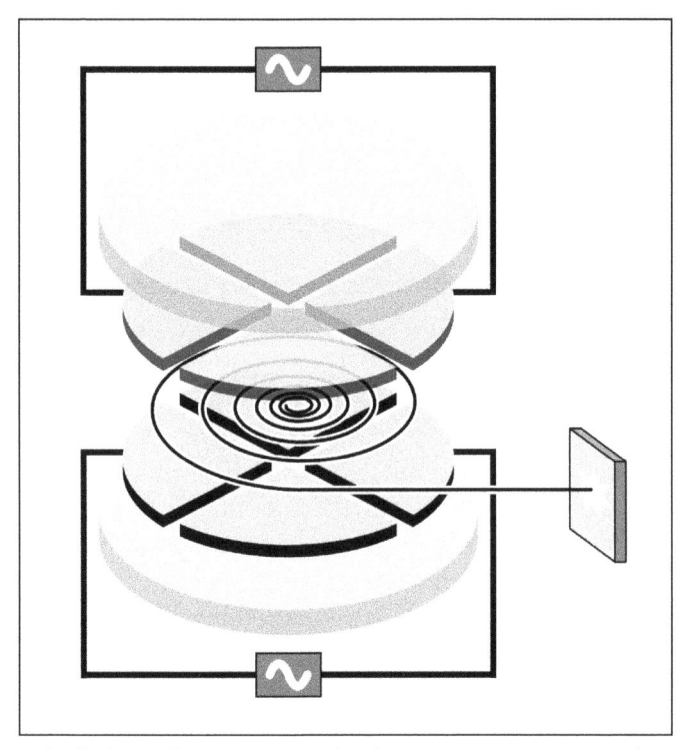

Figure 13. Cyclotron functioning principle scheme: protons generated at the centre of the cyclotron are accelerated in an area under vacuum by means of an electromagnetic field schematised by the four quadrants. When they reach their ideal speed, the protons are directed onto an external target containing the element to be transformed. The nuclear transformation reaction takes place in this target. When the production process reaches an end, the resulting radionuclide is isolated and purified in a radiochemistry cell that is close to the cyclotron unit.

may be solid, liquid or gaseous, transforming it into radioactive material. Unlike a reactor, a cyclotron is ideal when the radioisotope formed is an element different from the cold isotope serving as the target. Thus Fluorine 18, a positron emitter, is formed from Oxygen 18, a stable isotope available in the form of a water molecule, in liquid form, and Iodine 123, a γ emitter, is produced from Xenon 123, a stable gaseous isotope. At the end of the process, the radioactive element can easily be separated from the cold element that served as a target.

In a cyclotron, the particles are accelerated in a circular electromagnetic field under a high vacuum. When these particles reach a suitable speed, the beam is directed outside the field onto the target to be irradiated. The operation hardly lasts a few hours. The irradiated target is extracted from the cyclotron in order to be treated by radiochemists in such a way as to separate the radioactive elements from the residual cold matter. Products obtained via a cyclotron have an extremely high specific activity. In fact the products are virtually pure, but the quantity of material actually available can be counted in millionths of a milligram.

2 Generators

A generator is a small device containing a radioisotope of medium half-life which transforms slowly over time into an isotope displaying ideal characteristics for a nuclear medicine application. This "parent" isotope is fixed to a medium which acts as a filter and retains it. When washed with a saline solution, this medium releases the formed "daughter" radionuclide while the non-decayed parent still remains trapped. Technetium 99m, the most frequently used diagnosis isotope, is produced using a generator. This isotope with a half-life of 6 hours is formed by the breakdown of Molybdenum 99 with a half-life of 66 hours, and the fraction formed can be collected regularly each day for use on-site. The whole apparatus, the size of a five litre can, is composed simply of the filter column containing the Molybdenum, a pocket of saline solution and the tubing, the radioactive part being protected by fifteen kilograms of lead. A generator of this type is exhausted in about 2 weeks.

The generator is an ideal tool for the production of Technetium 99m because the latter is involved in more than two thirds of nuclear imaging techniques. The isotopes produced using generators are relatively limited. Nevertheless, there is at least one known example of a generator model for each radiation type on sale or partially in development, including alpha emitting isotopes.

3 Reactors

Most radionuclides used in nuclear medicine are artificial isotopes. The most usual method of production consists of bombarding a stable isotope (the target) with a flow of neutrons in a nuclear reactor. This neutron condenses with the existing atomic nucleus, thus creating an unstable isotope of the metal used as the target.

Molybdenum 99, used in the generator described above, may be formed by neutron bombardment of stable Molybdenum 98. The yield from this reaction is very low and Molybdenum 99 is extremely difficult to separate from Molybdenum 98. The mixture of cold and hot molybdenum could be used directly in the generator. This use remains limited because its specific activity, *i.e.* the ratio between hot and cold material, remains low. Expressed in radioactivity, specific activity seems very high, but brought down to a quantity level, the mixture contains less than one radioactive atom per several million cold atoms. For generators, it is therefore preferable to use molybdenum which is the product of uranium fission.

4 Fission Products

Many centres processing nuclear waste from fission reactions have access to large quantities of materials that are considered as being useless. Most of these are by-products of Uranium 235 fission used in nuclear power plants. They are stored while waiting for their complete breakdown, which may take thousands of years. Among these radionuclides, there is material that could be useful in nuclear medicine. It is possible, in some cases, to separate them from other isotopes. For example, Iodine 131 used in therapy is produced from this waste. Obviously, Molybdenum 99 is also available. It is a compound identical to that described in the neutron bombardment method, but much easier to separate from the other radionuclides.

III. The Production of Vectors and Ligands

Vectors, in other words organic molecules and synthetic peptides or antibodies, are traditionally produced by organic chemists and biologists. Their preparation must follow the production recommendations and constraints applicable to all pharmaceutical substances. The basic rules are set at an international level in the Good Manufacturing Practices (GMP) recommendations. No deviation from the synthesis and purification methods filed in the Marketing Authorisation dossier is allowed. This rule ensures that the final product's quality can be reproduced for each manufactured batch. Indeed, these rules are the same for both pharmaceuticals and radiopharmaceuticals.

A radioisotope can only very rarely be directly grafted onto the vector (covalent bond for halogens such as iodine or fluorine). During the course of development of the molecule, chemists must develop a method of attachment by trapping radioactive metal in an organic cage – a chelating agent – which ensures that the radionuclide is irreversibly attached to the vector. This stage of additional chemistry must obviously follow the same production constraints as does the production of the rest of the molecule.

IV. The Industrial Production of Radiopharmaceuticals

Unlike drugs manufactured in very large quantities, there is very little difference between the production of radiopharmaceuticals for clinical studies and the routine production for prescribing to patients. Everyday production corresponds exactly to the number of patients to be treated on the following day, or the day after, depending on the country, along with a few extra doses for testing the product's quality. Therefore, the production unit is often identical to the development unit, if they are not one and the same unit. Taking the short half-life of radiopharmaceuticals into consideration, their manufacture must be repeated each day. It can be compared to a "just in time" production process, subject to external constraints that may cause a real availability problem to patients (manufacturing interruption, transport strike, etc.).

When the production batch is greater than just a few units, it is not possible to check the quality of each unit individually. Several sample vials are reserved for product quality testing. The quality of the whole batch is guaranteed by the reproducibility of the manufacturing method. No deviation from the process or the specification is accepted. The raw starting materials must have been manufactured by a well-defined method, and themselves comply with pre-established specifications. Quality Assurance ensures that the methods of production and control are properly described and complied with. Quality Control is responsible for checking that the properties of the final product actually correspond to the specifications. These relate to both the composition of the product and its packaging, including the level of impurities and labelling, but also ensure that the product complies with sterility and apyrogenicity (the product must not be able to be held responsible for triggering a fever due to the presence of endotoxins of bacterial origin that cannot be destroyed during the sterilisation cycle, nor eliminated by filtration). The release of batches, therefore the authorisation for the day's production to be sold is the responsibility of the Managing Pharmacist, as in any other traditional pharmaceutical industry.

The safety aspect in the context of production has so far scarcely been touched on. It is well understood that the most significant difference between development and production concerns the quantities of radioactive substances being produced. Therefore, production areas are equipped and protected accordingly. Handling is carried out in shielded boxes, *i.e.* manufacturing cells provided with leaded windows and grabs. Technicians are protected by a layer of more than 10 cm of lead, and the cells contain all the material required for production. The staff who manufacture the drug work on one side of this production cell, the front, using remote manipulators (giant mobile grabs). They control their work through thick viewing windows made from Plexiglas or leaded glass. Operators who work on the other side are responsible for moving all the material required for the manufacturing in and out of these cells via large lead-lined airlocks. All the radioactive matter leaving these cells, including waste, is compulsorily confined in lead protective containers. By way of consequence, none of the operators is ever in contact with a radioactive product, except by accident. It is very difficult to automate this sort of production and each vial is produced one at a time, in an almost artisan fashion.

The routes travelled by the material are arranged in such a way that it is possible to define zones of increasing radioactivity, and in consequence, zones of increasing radiological risk. Thus, the front side technicians can never come into contact with the back side operators during the course of production. Only back side operators are working in an accidental contamination risk area. At any time, leaving these zones is controlled to the extreme. Each operator carries a radioactivity detector and must also go through more than one check before leaving his/her place of work. In principle, radioactivity can only leave the site in leakproof leaded containers, whether it is a question of drug vials or of waste in sealed drums.

V. Transport and Logistics

Several hundred thousand consignments per year, that is several thousand per day, leave production centres for delivery to their final destinations in as short a time as possible, but above all without the slightest incident. No means of transport is excluded. Nevertheless, road and air are favoured. At the end of the production day, dozens of lorries wait to be loaded outside the despatch warehouse's exit. Their arrival and departure times depend on the final delivery time or the plane's take-off time. Generally speaking, the destination hospital should receive the delivery within twelve hours. Ideally, the active product is delivered for use the next morning as the hospital's nuclear medicine department opens its doors.

A number of parameters may interfere with the regular operation of these logistics and nevertheless imperil delivery deadlines: the state of the traffic, customs checks, acceptance of the deliveries by the pilots in command, etc.

Radioactive products of all types are packed in special containers resistant to falling, crushing, fire, etc. This reduces the risk of accidental contamination linked to the transport system. Incidents during transportation are extremely rare. A frequency rate in the order of 1 to 2 incidents per 100,000 consignments seems completely insignificant, but is nevertheless unacceptable and must be considered on a case by case basis. Each event has led to the analysis of both the incident and the risk. This resulted in further container change and general process improvement and, therefore, to higher safety levels.

With the increased risk of terrorism, the hijacking of means of transport for ill-intentioned ends has been subject to deeper analysis by the state security services. Radiopharmaceutical products are consignments that are not very effective for such actions, given their half-life and activity. The risk associated with their dissemination or reuse is almost nil. At the very most, there would be a psychological impact on the public and journalists, due to a lack of knowledge of this field.

VI. Radiopharmacies

All products dispatched to a hospital are not delivered ready to use. Some handling and checks are necessary, if only to regulate the dose that is going to be injected. Certain products even require final on-site production. This is the case for all products using technetium delivered in "cold kits" and which must be labelled on-site with technetium produced using a generator. In the nuclear medicine department, it is the radiopharmacist's job to complete this preparation. He/she becomes the final producer. Complying with the same safety and quality rules as those imposed on the industry, radiopharmacists make up the various solutions to be injected each morning. Cold kits contain all the reagents that enable the formation of the expected radiopharmaceutical once they are combined with the right quantity of technetium from the generator. Normally, this stage involves simple mixing, and possible stirring over low heat, followed by quality checking, but every single step must follow a very strict protocol. The radiopharmacist becomes the local guarantor of the reproducibility of this last stage and the quality of the final product.

The quantities of matter used are so small that traces of air are sufficient to alter the result. All materials used are sterile, but the final product's sterility also depends on the experience of the person handling it. Finally, each handling procedure is different, to the extent that it must be adapted to the quantity and concentration of technetium from the generator.

Radiopharmacists' work is not limited to reconstituting technetium kits, but as the use of these products increases, this represents the greater part of their work. They must, however, be capable of meeting all the demands of the nuclear medicine department and, given the diversity of isotopes used, their field of knowledge is relatively vast. The recent arrival of therapy kits diversifies as well as complicates their field of action. In particular, not all departments are currently equipped to handle activities on such a large scale, and some laboratories need to be improved in terms of staff protection level. Specific authorisation needs to be requested, department by department. In some countries, and especially in Europe, the authorities have recently given the resources to set up a system for auditing and monitoring these hospital units.

The Patient in Hospital

Nuclear medicine departments located within a hospital are hardly any different from other departments from a structural point of view, and, if they are not really paying attention, patients can leave without ever having noticed all the security aspects that have been implemented. It often happens that cardiac patients come out of scintigraphy without realising that they were injected with a radioactive substance. For the uninitiated, it is even more difficult to distinguish a SPECT camera from an MRI one. And inside a hospital room, no-one is capable of estimating the thickness of the walls and their lead content. However, signs in the form of clover leaves as a reminder of the presence of radioactivity are everywhere, as are the various protective elements for staff (airlocks, windows, walls, lead-shielded syringes, individual rooms, etc.).

The procedure itself is hardly any more restricting than the injection of a contrast product for an X-ray examination. The pain will be limited to that caused by the needle. In order to reassure patients, nuclear medicine departments have short explanatory leaflets available concerning the procedure and the risks, documents that a nuclear physician will not hesitate to explain and further detail.

However, as soon as they have been given the injection, patients themselves become a source of radioactivity up to the point where the substance is completely eliminated, either because of decay, or by biological elimination. The documentation supplied by the medical department will also explain the few rules to be followed during this period. In particular, as most of the substance will be eliminated via the urine, patients will be asked to use special toilets allotted for this purpose in the hospital in order not to contaminate the environment. The hospital is equipped with special tanks that will only be emptied when the decay of this waste has reached the allowed threshold. This is particularly true in the case of therapy. For imaging, the quantities injected are so small that there is no reason to keep patients in hospital. It is actually preferable that they should return home, because the most exposed people are the nurses and physicians who are in daily contact with all these patients. The dose transmitted by patients to their entourage is almost negligible as compared with ambient radioactivity and, as a general rule, no special precautions are required at home.

Radioactivity does however exist for a few hours to several days, and remains detectable even at very small doses. Anecdotal cases have caused a sensation in the newspapers. For example, urinating into a river close to a nuclear power station in the hours following scintigraphy may set off the site's radioactivity detectors, as these are sensitive enough to pick up this minor change in the water's radioactivity level.

At this stage, it would be useful to remember that the radionuclides used in nuclear medicine all have very short half-lives, and that the disappearance of radioactivity occurs quite "naturally" in the hours and days following injection or absorption.

VII. Nuclear Medicine Centres in the World

The world market for imaging agents, all modalities included, represented in 2002 a turnover of about 4.5 billions euros. Among the 800 millions images which were taken during this same year, about 120 millions required the injection of a product. About 28 millions patients benefited from nuclear medicine examination using radioactive products, for a total cost of 1.7 billions euros.

THE PRODUCTION OF RADIOPHARMACEUTICALS   141

FROM THE RADIONUCLIDE TO THE PATIENT

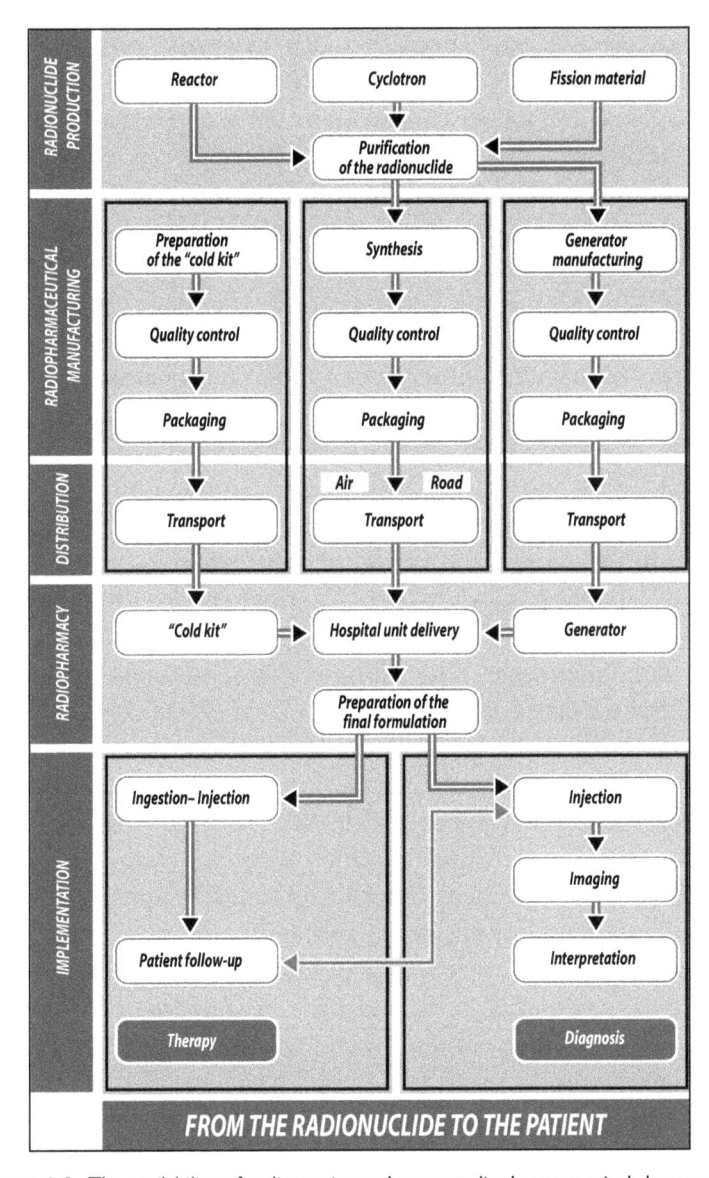

Figure 14. The availability of a diagnosis or therapy radiopharmaceutical drug requires the initial production of a radionuclide that will be transformed into a radiopharmaceutical in an industrial environment that complies with all pharmaceutical quality criteria and nuclear safety rules. It will then be packaged and transported to the hospital following these same quality and safety standards. The constraints of short half-lives add an extra difficulty, as the process must be scrupulously respected for risk of loosing the final product as a consequence of delay.

There are only few companies that share the contrast agents (X-ray, MRI and ultrasound) and nuclear medicine products world market. Seven companies control about 90% of the market: Amersham Health (recently taken over by General Electric), Bracco, Bristol-Myers Squibb Medical Imaging (formerly Dupont), CIS bio international (subsidiary from IRE and IBA), Tyco Healthcare (Mallinckrodt Imaging), Schering and Guerbet. The first five companies are also, in no particular order, the major producers of radiopharmaceuticals. Several smaller companies have become specialised in a limited number of products, sometimes only one, FDG, but do not compete in terms of turnover with the companies showing the above-mentioned figures: AAA, Alliance, Cardinal Health, Cyclopharma, IBA Molecular, Immunomedics, Jason, Siemens-CTI with Petnet, etc. In fact, the majority of the FDG manufacturing centres are run by one-site company structures, or are state laboratories, while SPECT products are distributed via local radiopharmacies which are sometimes associated in a network. These radiopharmacies are supplied with radiopharmaceuticals from the above-listed larger companies.

Public and private nuclear medicine centres need an official authorisation to handle radioactive products. In developed countries, the ratio of nuclear medicine units is about one centre per 250,000 inhabitants. These centres are located within university hospitals, cancer treatment centres and public hospitals or private clinics. They must be authorised to handle non-sealed radioactive sources (solutions for injection or capsules for oral administration), and are equipped with dedicated beds in shielded rooms (leaded walls, specific waste treatment units, etc.) In average, each nuclear medicine unit has access to two SPECT cameras.

PET imaging centres are nuclear medicine units equipped with one or more dedicated PET cameras. Each centre is linked to a Fluorine 18 manufacturing unit (cyclotron) located on site or at less than two to three hours driving distance. In a well equipped country (*e.g.* USA, Germany, France, etc.) cyclotrons are located in such a way that the whole territory can be supplied with FDG. By way of consequence, in those countries the limitation in number of diagnosed patients is rather linked to the number of cameras, and not to the number of cyclotrons. Among the limited number of well equipped countries, it seems today that only the USA have a sufficient number of cameras to cater for every single patient with a need. In all other countries, waiting times are usually of several weeks for a scan, if available at all. Surprisingly, by the end of 2006, countries such as Great Britain or Canada are only just starting to equip themselves. Of course, this technology's development in a specific country is a highly political decision that must be integrated in the overall health management structure (cost and reimbursement), as it has a serious impact on the country health budget.

By the end of 2005, the five largest European countries (France, Germany, Italy, Spain and UK) had implemented more than 1,200 nuclear medicine centres in total, that ran about 2,700 SPECT cameras. In Europe, Germany is twice as equipped as all other countries, and on a world scale, Japan alone runs more than 1,700 cameras in about 1,300 centres. In terms of PET units, the level of equipment depends on each country's priorities. The geographic areas can be split into three groups: the starters that are very well equipped, almost over-equipped (USA, Japan, Germany, Belgium, Northern Italy), the countries that have now started and are in a process of catching up with the first group (France, Spain, Turkey), and the countries that have only just decided to invest in this technology (Canada, Brazil, Portugal, Poland, Hungary, Morocco, Slovakia, UK, etc., but also China and India), or are still under discussion to decide when they will start (Algeria, Tunisia, CIS countries, South America, etc.) In 2005, Europe was equipped with about 120 cyclotrons and 300 PET cameras among which about 200 were already installed in the five largest European countries. The USA alone have as many as 800 PET cameras. On the other hand, the 15 Eastern European countries currently own 250 nuclear centres between themselves, with about 400 SPECT cameras, but only 9 PET units. As these figures are obviously increasing fast, patients will benefit from improved access to this technology in the very near future.

Most of the countries have reached a consensus on the number of public PET cameras to be implemented on their territory. By 2000, a common accepted figure was one camera per million inhabitants. This ratio is slowly increasing and is closer to 1 per 800,000 in 2006, with a trend to become 1 per half million by 2012, as a consequence of the potential new indications in cardiology and neurology. In the USA, the number of cameras is already above this ratio and the figure should double again between 2006 and 2010. But this country is an exception and all other countries (except Belgium) are still well below this ratio.

Nuclear medicine is a speciality with which the public is not well acquainted. This is mainly due to the fact that civil use of radioactivity, even for medical use, has not benefited from advertising. Patients know this technology mainly through a single modality, myocardial perfusion scintigraphy. But even in this case, very few patients associate this technology with radioactivity. Contrary to MRI (Magnetic Resonance Imaging), NMR (or Nuclear Magnetic Resonance) is another non-radioactive imaging technology where physicians did not consider the suppression of the word nuclear to be necessary. Apparently the use of this wording seemed not to be detrimental to this technology.

Nuclear physicians themselves do not see an advantage in advertising for their technology. Except for the private centres which need to be profitable, nuclear physicians do not have to look for new customers as their waiting rooms never empty. Patients experience this level of saturation through the long waiting time for any appointment.

Constraints for patients are clearly linked to the limitation of equipment and the lack of trained staff. In order to accelerate the decisional process, they are often offered to switch to another imaging modality, which obviously does not provide the same level of diagnosis information.

Beside the lack of equipment, a problem that can be solved within a decade, one must take into account the fact that well-trained nuclear physicians will be needed in every country, and that this will take much longer to achieve than simply buying a camera, or even building a cyclotron. There is an urgent need for young nuclear physicians, partly because most of the currently handling physicians are to retire soon. It seems that PET will bring a new lease of life to this specialty, with a whole new generation of physicians getting involved in nuclear medicine.

Summary

Radiopharmaceutical products are constituted from a non-radioactive fraction, the **vector**, combined with a radionuclide. Vectors are prepared the traditional way, by any method of chemical or biological synthesis imaginable.

Radionuclides can be obtained in a variety of manners:
– direct synthesis using a dedicated **cyclotron**;
– extraction from a **generator**, itself loaded with a parent isotope originating from a cyclotron or reactor;
– purification of radioactive substances from a **reactor**;
– reuse of nuclear waste from another source (fission).

Before they are put on the market, radiopharmaceuticals are subject to very strict quality testing, in fact identical to that imposed for all drugs.

The complex nature of radiopharmaceutical use is above all linked to the half-life of the radionuclides, and requires special infrastructure and logistics which become even more complex when fluorinated products must be distributed with a half-life of less than 2 hours.

The infrastructure required for administering radiopharmaceuticals in a hospital environment is just as sophisticated. All these environments are subject to regular monitoring, in order to preserve biological as well as nuclear safety, and necessitate highly qualified staff.

The future of the technology looks promising but will only develop as and when new cameras and treatment rooms are installed.

CHAPTER IX

Future Prospects

The arrival of the first effective targeted radiotherapy treatments brought a new dimension to nuclear medicine. With the exception of their use in radio-synoviorthesis, in thyroid cancer treatment (highly efficient), and in reducing cancer pain (limited use), nuclear medicine tools were, until now, more widely used in diagnosis imaging. During the course of recent years, the arrival of hybrid imaging tools and the combining of different, but complementary, diagnosis techniques, has given new impetus to this technology allowing better pathology definition, better patient monitoring, and, above all, better treatment direction. A new avenue is also opening up with the arrival of individualised medicine in which imaging plays a key role. Verifying a therapy's suitability for a sub-group of patients becomes much easier with these new tools, via an easier patient selection which in turn ensures increased treatment effectiveness.

I. Hybrid Imaging Tools

The increased difficulty of locating small tumours via scintigraphy can be got round easily by combining nuclear imaging tools with conventional tools. Since the summer of 2000, medical imaging devices manufacturers provide physicians with new tools allowing them to simultaneously obtain a three dimensional image which is both functional and anatomical. In previous years, researchers had tried to compensate for this lack of resolution in scintigraphic images by acquiring two consecutive scans, one after injecting a radioactive tracer, and the other using X-ray or magnetic resonance imaging techniques. Although computerised data processing has taken a considerable leap forward, the complexity of superimposing images of patients who were moved from one table to another between the two scans limited this technique to research applications, and remained totally disproportionate for routine implementation. The arrival of mixed SPECT/CT or PET/CT scanners can be considered to be a real revolution in the field. Hybrid SPECT/MRI or PET/MRI tools are under development and could really bring further advantages.

Today four large companies dominate the world market in medical imaging equipment: General Electric (with a dedicated imaging subgroup called General Electric Medical System) which is the leading world manufacturer with more than a quarter of the market and is working in all imaging fields, Philips (actually Philips Medical Systems), the leader in digital technology, which owns 15% of the world imaging market and is above all leader in the field of MRI, Siemens, the leading European manufacturer and covering 19% of the world market, and Toshiba, more oriented towards networks specialised in X-ray equipment. Smaller companies which were trying to compete with the four giants in special niches were taken over one after the other in the past years. In October 2003, General Electric took control of Amersham Health, the world market leader for imaging products, thus creating the first group capable of offering both the tools and the imaging products. By acquiring the company CTI (cyclotron manufacturing) with its PET radiopharmacies network, Siemens entered in the same merger strategy.

II. Individualised Medication

The doctor's role consists of identifying the pathology, of determining its origin and of fighting the sickness in order to heal the patient, or at least to provide some relief. Physicians are capable of making a diagnosis on the basis of their knowledge and experience, using tools that have been provided for this purpose. These tools have evolved over the year and provide increasingly precise analyses. The increase in knowledge has contributed to the creation of specialties that are more and more specific, one person alone no longer being capable of knowing everything about human biology and medicine.

Today, for each diagnosis, and therefore for each pathology, there is a corresponding specific treatment. Illnesses themselves have been divided into sub-classes, each with its own therapeutic protocol.

Nevertheless, although these protocols have been able to demonstrate their effectiveness for large groups, there are certain patients who do not respond to an otherwise ideal treatment.

These patients' lack of response to a given treatment is linked to specific types of resistance or an absence of responsiveness inherent to their genetic heritage. In the same way that there are populations who transmit disorders from generation to generation, other types of individuals hand down their resistance when facing certain external attacks. Metabolic processes also differ from one population group to another, and some patients transform drugs which were administered to them faster than others, thus making them less effective. These individual differences seem to be directly linked to a specific genetic heritage as well.

A cancer patient resisting an initial treatment could be offered a second treatment after several weeks of inefficient therapy, and even a third later on again. During all this time, the cancer continues to develop and when an appropriate treatment is eventually administered, it sometimes is too late for it to be effective. It becomes urgent, therefore, to find a method that can determine well in advance whether patients are going to respond positively or not to a specific treatment, even before it is administered. This is the role of theranostic tests. This word was coined by combining the terms 'therapy' and 'diagnostic' (diagnosis). It could be explained as being a method of diagnosis and selection of a patient (and not a disease) with a view to a particular therapy. While the objective of a diagnosis is to identify the illness from which the patient is suffering, theranostics must determine the treatment best suited to heal this individual as quickly as possible.

The concept goes even further: depending on the response to the theranostic test, it should be possible to predict the chances of healing with a specific treatment, to calculate the exact doses required and to follow up the patient in the course of remission. Treatment then becomes individualised because it is directly linked to the patient's genetic heritage.

In 1999, herceptin, a drug used in breast cancer therapy, was the first example of a treatment based on this concept. The American administration only authorised the sale of this product provided it should only be applied to patients who demonstrated the presence of a herceptin receptor. In this case alone, treatment was likely to be effective. This is still a long way yet from the selection of patients on the basis of their genetic heritage, but this example is sufficient to show that by limiting treatment to a sub-population, its effectiveness is almost guaranteed. Conversely, non-responders to the test should seek other therapeutic methods without wasting time with a protocol including herceptin.

The pharmaceutical industry is in general making slow progress in this field, for the simple reason that selecting sub-groups of patients inevitably leads to a reduction in the population to be treated, and therefore to a reduction in turnover. The investment being the same, whatever the size of the group of patients being targeted, this is not the direction in which the major groups will invest funds. On the other hand, small companies could more easily launch themselves into this battle, under the guise of so-called orphan treatments (*see below*), which are less costly to develop.

Nuclear medicine has an important role to play in this field. Before even targeting the selection of patients on the basis of their genetic heritage – their DNA or RNA could of course be imaged by labelled substances – it is possible to envisage the display of the distribution of small quantities of substances in the sites to be treated. If patients do not show an ideal diagnosis drug distribution, it is not worth inflicting an ineffective treatment on them. As nuclear imaging allows the selection of a sub-group of patients, this technology could also be considered as a theranostic method. Combining theranostic products with diagnosis and therapeutic products, in the context of legislation on orphan drugs, opens a fast track for nuclear medicine. A pharmaceutical company ready to invest in this area remains to be found. The new conglomerate formed in 2003 by the purchase of Amersham Health (manufacturer of imaging products) by General Electric (producer of imaging equipment) seems to be heading in this promising direction.

III. Orphan Diseases and Orphan Drugs

Taking the high cost of developing a new drug into consideration, and the less than philanthropic character of the pharmaceutical companies (and above all of their shareholders), legislators have had to set up a system that encourages some laboratories to take an interest in less profitable diseases. These so-called orphan diseases only affect a very small part of the population, and allocating financial resources to this field is only of interest if there is a guarantee of a minimum return on the investment. For several years now, a specific law gives American companies certain advantages such as an exclusive market right if they develop a drug corresponding to the definition of an orphan drug. More recently, a similar law has been voted in Europe and has been implemented in some countries since as early as January 2000.

A drug becomes orphan if it is intended for the diagnosis or the prevention or the treatment of a disease affecting less than five persons in ten thousand within the European Community. It must be developed for treating a serious disease, or one that is incapacitating, and it must also be very unlikely that it would be marketed without incentive measures. It is obvious that this drug must in addition prove to be superior to any other known available treatment. In some countries, this definition also covers diseases that are very rare in the country itself, but which affect millions of people in Africa, and for which research efforts remain limited due to a lack of financial capacity from concerned customers, patients and countries.

Taking the development costs of a drug into account, therapy for an orphan disease is extremely costly anyway, and politicians' support becomes a key issue.

Dossiers of new drugs filed in the context of this procedure enjoy accelerated processing, reduced or even no taxes and exclusive marketing rights for 10 years. In addition, the contents of the dossier are reduced as compared with the dossier for a drug targeting a wider population. In Europe, the dossier is centralised, in other words its application is immediately extended to all member states. Orphan drugs may benefit from other incentive measures taken at local levels, in order to promote research, development and marketing, and some Health Agencies have undertaken to assist companies in developing their product and in preparing their dossier. More particularly, research measures in favour of small to medium sized companies have been set up.

Initial applications processed by the European Medicines Evaluation Agency since these texts came into force have to do with drugs intended for genetic diseases leading to early death, the treatment of cystic fibrosis, certain rare cancers, diseases for which the only therapeutic option lies in a transplant, and above all cancers for which all known treatments have failed.

These last examples become significant in the context of metabolic radiotherapy. Not only is this technology in the process of demonstrating that, in certain circumstances, it is capable of compensating for the deficiencies of traditional therapies, but the new legislation should allow innovative products to be put on the market faster, as all the concerned indications fall within the context of orphan diseases.

It is true that, to date, the therapeutic application of nuclear medicine techniques is limited by the environment in which they have to be conducted. As long as it is possible to demonstrate on a larger scale that this technique can be a substitute for chemotherapy for example, it is obvious that the applications will exceed the framework fixed by the legislation concerning orphan diseases. Legislators have already foreseen these circumstances, and allow backtracking and re-entering the framework of development of a traditional drug. Taking a widened market into consideration, a company is assumed to receive greater profits than initial forecasts. It will therefore have the necessary financial resources for additional studies, which will in turn minimise the risks to patients, as these increase proportionally with the growth of the population.

This law should promote the development of nuclear medicine products by small companies, which do not target the world market in its widest sense.

IV. Ethical and Regulatory Limitations

1 Regulation and Administration

Radiopharmaceutical products are drugs that combine both the constraints of the pharmaceutical domain and those of the nuclear domain. Firstly it is necessary to develop, produce and make available, both to physicians and patients, a product the effectiveness and pharmaceutical and biological quality of which is guaranteed. Secondly, the production, distribution and administration of a radioactive substance must be guaranteed, while protecting the environment and staff coming into indirect contact with it. In consequence, two distinct administrative authorities usually have, under the control of two separate ministries, the duty and power to monitor compliance with specific rules at all times. This monitoring does not just apply to developers, producers and transporters, but also to users, that is to say to hospital physicians and radiopharmacists. In particular, these authorities are responsible for checking the suitability of the relationship between the licences they grant and the training given to staff in using radioisotopes, the latter sometimes being managed by a third ministry. If the legislation concerning pharmaceutical products is constraining, that concerning radiopharmaceuticals is even more so.

Side Effects and Toxicity

Each time a drug is administered, and not just in the case of radiopharmaceuticals, there potentially exists an associated danger for patients. Physicians who are aware of the limits of the products they administer must minimise this risk while pressing on with the treatment in such a way as to promote healing. Most drugs are extremely harmful products, but are used in doses where this toxicity is not evident. In certain circumstances, some side effects which are in reality the visible side of these toxic effects, are tolerated. Chemotherapy, a treatment that is extremely aggressive for the whole organism, is unfortunately only efficient at doses that also create significant side effects such as nausea and vomiting, hair loss and gastric problems, and above all bone marrow toxicity. Reducing a dose in order to minimise the side effects only reduces its effectiveness, and is therefore more risky to the patient. While waiting for even more specific and less traumatic chemotherapies, many palliative treatments for these side effects have been developed. At this level of therapeutic need, the side effects observed are accepted by physicians and tolerated by patients.

In the case of radiopharmaceuticals, the risk is doubled. Patients firstly suffer from a vector's toxicity equivalent to that described for pharmaceutical products, and secondly from the effects associated with radiotoxicity. Taking the extremely low quantities of active materials injected in patients into consideration, the toxicity of the vector plays no role, except perhaps in allergic phenomena. As a consequence, developers of new radiopharmaceuticals have the means to accelerate their research process without dwelling too much on toxicity problems inherent to the vector. They are even able to use, as a base, molecules which are rejected by the pharmaceutical industry because of their high level of toxicity.

In all cases, radiotoxicity will need to be taken into consideration as something of first importance. Researchers are increasingly concentrating on the same small group of radionuclides, the radiotoxic characteristics of which are starting to be well-established or at least, where the doses which should not be exceeded are known.

As with chemotherapy, higher doses have significant consequences for the individual and also lead to modifications in blood count, even to bone marrow failure. On the other hand, external or metabolic radiotherapy is only rarely accompanied by side effects. At the very most, a certain level of digestive troubles has been observed, as well as a burning sensation at the injection point. Therapeutic nuclear medicine has the great advantage of displaying very limited drug-linked morbidity for patients. For diagnosis analysis, hospitalisation is rarely required, and is not linked to the imaging method, but to the treatment or monitoring of the patient.

Dosage and Indications Extensions

The quantity of radiopharmaceutical as delivered to the physician for injection is usually slightly higher than the quantity to be administered to the patient. Due to decay, the dose on leaving the factory or the radiopharmacy is much stronger and takes account of the theoretical injection time. The nuclear physician will have to calibrate the dose for the time of application to ensure that the actual prescribed quantity will be fully injected into the patient. A possible dosing or handling error on the part of the radiopharmacist or physician must also be taken into consideration.

As a precautionary measure, but without any evidence to support it, children and pregnant women are normally excluded from diagnosis procedures involving radionuclides. Nevertheless, in extreme situations, children can undergo radiotherapy. If it is not too late, it is preferable to offer pregnant women suffering from cancer an abortion before treatment. In fact, these questions, raised by physicians, are no different, whether they concern a child or a pregnant woman undergoing chemotherapy.

Legislation concerning annual dosage limits for injected radioactive substances does not apply to patients. It is obvious that any examination must be justified for a certain indication, as well as for the equipment used. The expected benefit must exceed the long term risk by a wide margin. It must also particularly take into consideration the possible appearance of new cancers. This is one of the reasons for which therapeutic treatments are generally only intended for patients for whom there is no other alternative.

However, limiting needless radiation remains a priority. The choice of dose delivered must remain sufficient so as not to compromise the diagnosis result or the therapy's effectiveness. As with chemotherapy, radiotherapists and nuclear physicians will work at the limits of toxicity with a concern for maximum effectiveness and minimum risk. To this end, they are ready to accept some unavoidable transient side effects.

V. Politics and Legislation

Radiologists and nuclear physicians both need a certain level of autonomy in their respective sectors. Each specialist's field of competence still needs to be clearly defined, and the situation gets even more complicated when budget and funding become part of the discussion. In fact, radiologists, established for longer and responsible for X-ray and Magnetic Resonance Imaging, could claim to cover the whole diagnosis sector. Nuclear physicians on the other hand, specialists in a very specific field and responsible for all the consequences of using radioactive products, consider themselves more as practitioners who evaluate a disease's progression than just technicians, responsible for a specific tool. Unfortunately their operating budget is one of the lowest in the hospital, and the number of hospitals equipped with a nuclear medicine department is also very low. A further complication is added with the arrival of metabolic radiotherapy products which could be part of the radiotherapist's field. Unfortunately the latter, even though they are nuclear specialists, are only authorised to apply radiation from an internal or external, but sealed, source. As seen so far, as metabolic radiotherapy only has specific applications in certain fields such as oncology or rheumatology, it is prescribed by oncologists, haematologists or rheumatologists who are sometimes considering nuclear medicine to be a service department.

A nuclear physician does not have the training nor the experience of an oncologist. An oncologist does not have knowledge of radioactive material handling. A radiotherapist is not authorised to inject a radioactive product, and a radiologist is only involved in diagnosis techniques. As a consequence, all of them must work together, as a team. The hospital departments in which nuclear medicine has made the most progress are those where both nuclear physicians and oncologists manage to get on well with each other. Indeed, in this way both can find their place and be useful to one another, without attempting to manage the other's department.

A nuclear physician's annual budget is provided to cover a certain number of tests in the area of cardiology and oncology. It would be spent within a few weeks if it was also intended to cover the costs of therapy products. Oncologists and haematologists have the most expensive drugs available for use in chemotherapy and could in the future make substantial savings by replacing certain chemotherapy treatments with radiotherapy treatments, but are not yet sufficiently trained in, or rather informed about, these techniques. A few more years are required in order to improve budget allocation at hospital level, but this will only happen if the health authorities set new rules which would be beneficial to both patients and the state budget.

On the basis of budget limitations affecting health spending and the reorganisation of departments, the only alternative for all these specialists consists of implementing real co-operation and multidisciplinary team work. Within a few years, centres specialising in metabolic radiotherapy, and enjoying very good reputations, will appear. These centres will have succeeded in creating an effective team for the treatment of cancers using all the new techniques, based around a unifying and charismatic manager, whether this manager is an oncologist, radiologist, radiotherapist, nuclear physician or a hospital director.

Germany, for example, is also struggling with a growing number of medical specialities which only compartmentalise hospitals and limit the success of team work. The European administration is taking a closer interest in this problem, and it would not be surprising if texts were to appear shortly at a European level, to regulate the division by specialist sectors of the skills required in a hospital.

VI. The Future

Since it really demonstrated its effectiveness 50 years ago, nuclear medicine, and more particularly radiopharmaceuticals, progressed in stages. The 50's, up to the middle of the 60's, showed the therapeutic efficacy of Iodine 131 and Phosphorus 32 used in the simple chemical form of salts. At the same time, Technetium 99m and certain derivatives made their appearance, in particular pyrophosphates allowing the first images to be produced. During the 15 years that followed, other Technetium 99m derivatives were developed in the form of organic complexes, colloids, macro-aggregates and salts, which produced images of specific organs and tissues. From the 80's onwards, technetium chemistry evolved in such a way that substances which could be used in the perfusion of specific tissues could help observe the functioning of certain organs, such as the heart, the brain and the kidneys in particular. In the 90's, research turned towards targeting groups of similar cells. Through this new approach, substances concentrating in specific tissues were identified, and thus a certain class of cells could be located. It is only very recently that chemists suggested replacing imaging isotopes with therapy isotopes on complex molecules in order to destroy the cells to which these molecules bind. It will take 10 years to prove the effectiveness of these products. We are only at the beginning of this therapeutic revolution. The first years of this century have seen a new class of therapeutic products appear, which are showing real clinical effectiveness in the treatment of cancers. Other products are in the course of development, at different stages, and could appear on the market within the next decade. The development of this technology, which requires specific equipment, depends on two key factors: firstly, the financial and human resources granted to this field of medicine for its development with patients, and secondly the reception it is given by doctors specialising in oncology (competitive aspect), politicians (demagogy aspect) and the public (safety aspect). Finally, it is the patients themselves who are the direct beneficiaries of the technology, and they are the only driving force (such as lobbying groups) who will push for a faster development of this field in the short term. Because of the high costs, of the limited number of approved handling centres, and above all of the lack of knowledge of this technology, metabolic radiotherapy has been confined to the role of a last hope. Clinical studies in progress involving the new products are only initially intended for patients who do not respond to "standard" therapies (chemotherapy and external radiotherapy). Results achieved in recent years, and the availability on the market of new and effective products filling the gaps left by chemotherapy, will doubtless stimulate developments in this field. Nevertheless, more time is still required before a new range of products appears on the market. Step by step, metabolic radiotherapy will move from being only used to treat incurable cancers and end-of-life patients to being a second-line treatment. Then, if it proves its worth, it could even replace certain chemotherapy protocols. Each stage must be demonstrated. A study with patients suffering from non-Hodgkin's lymphoma is already underway in the United States. The patients involved in this study are being offered treatment using a radioactive product, not having previously undergone chemotherapy. The results are not expected before 2007.

In oncology, the reference criterion is survival. This criterion is easy to measure when physicians limit their interest in patients who have a limited survival time. They just need to count the patients who are still alive 2 or 3 years after the treatment and compare this number with the control group in order to prove the treatment's effectiveness. If the study consists of treating patients whose life expectancy is in excess of a year, it is no longer 2 years they have to wait, but at least 5 before giving their verdict. In addition, these studies will always be limited to a sub-group of patients suffering from a well-defined cancer. Individualised therapy is becoming increasingly evident.

These studies have not yet started and the first patients to benefit from them are those who have agreed to take part in them. If one of these takes place in 2006, commercial authorisation for the product tested will not be given before 2013, or even 2015. At the same time, it is to be hoped that other products will pass the initial stage and will be marketed for other indications.

Truly effective metabolic radiotherapy drugs are only just beginning to appear on the market. They are the forerunners of products on which great hopes are based, and they will soon compete with chemotherapy products to the patient's benefit. The examples described in previous chapters essentially concentrate on labelled antibodies in order to better describe the advantages, but also the constraints, associated with this technology. Other avenues should lead to the introduction of new products and new technologies in hospitals within the next 10 years.

Glossary

Absorbed dose: quantity of energy transferred to a substance per kilogram during absorption of radiation, expressed in grays.

Acquisition: recording of all the radiation accumulated over the course of a predetermined period of time in order to obtain an image.

Activity (radiological): value representing the number of disintegrations per second from a radioactive source, expressed in becquerels.

Affinity: property of a substance to bind to a receptor; it is a measurement of the strength of the binding.

ALARA: As Low As Reasonably Achievable. Radiation protection policy related to the management of staff working in an ionising radiation environment.

Alpha (α) (alpha radiation): a particle emitted by a radioisotope and formed from a nucleus of helium containing two protons and two neutrons with potential therapeutic uses due to its strong ionising power.

Auger (electrons): low energy electrons emitted from the surface of an atom, but endowed with ionising power at a short distance and therefore usable in therapy.

Becquerel (Bq): unit of radioactivity equal to one disintegration per second. The becquerel replaces the former curie unit, one curie being the equivalent of 37 billions becquerels.

Beta-minus (β^-) (beta-minus radiation): a particle emitted by a radioisotope and formed from a negatively charged electron, usable in therapy due to its destructive potential.

Beta-plus (β^+) (beta-plus radiation): a particle emitted by a radioisotope and formed from a positively charged electron (positron), an unstable anti-electron, which when it meets a negatively charged electron is annihilated to emit two gamma photons which move in exactly opposite directions, and are therefore usable in imaging. This radiation could possibly be used in therapy.

Biological half-life: time period at the end of which a cell or tissue has eliminated half the quantity of a molecule present by a biological metabolism mechanism followed by excretion.

Brachytherapy: method of internal irradiation by the temporary or permanent introduction of radioactive implants. Examples: radioactive seeds marked with Iodine 125 in prostate tumours, iridium wires in breast tumours, Phosphorus 32 patches.

Cold kit: non-radioactive precursor of a radiopharmaceutical containing all the elements that enable this medication to be reconstituted instantaneously, simply by adding a radionuclide solution.

Computerized Tomography (CT): cross-sectional imaging allowing three-dimensional reconstruction.

Contamination: physical contact leaving a deposit of radioactive material on a surface, matter or person. The contaminated person is irradiated as long as the active matter has not been eliminated or the radioactivity has not fully decayed naturally.

CT: abbreviation of Computerized Tomography.

Curie (Ci): one curie equates to the radioactivity emitted by one gram of pure Radium 226, one of the first natural radioactive materials available and isolated at the beginning of the last century. In principle, this unit should no longer be used because it was replaced by the becquerel (*see entry for this word*) in the 80's.

Curietherapy: *see Internal radiotherapy.*

Decay: reduction in the degree of radioactivity over the course of time.

Dosimetry: the study and measurement of absorbed radiation.

Effective dose: the equivalent dose corrected by the weighting coefficient relating to the irradiated tissue (0.05 for the thyroid, 1 for the whole body) expressed in sieverts.

Effective half-life: radioactive half-life corrected by the biological half-life. With it, the practitioner can estimate how long a radioactive substance that has been ingested or injected will take before generating an effect on the organism (or a certain type of cell or tissue).

EMEA: European Medicines Evaluation Agency, European Health Authority, centralised in London.

Equivalent dose or Dose equivalent: absorbed dose corrected by a weighting coefficient relating to the radiation (1 for X, beta and gamma rays, 20 for alpha rays), expressed in sieverts. This is a value used in radiation protection to take account of the difference in biological effect of the various types of radiation.

External radiotherapy: method of therapy by irradiation using a source external to the patient (cobalt therapy). Domain of the radiotherapist.

FDA: Food and Drug Administration, American Health Authority.

FDG (Fludeoxyglucose): substance labelled with Fluorine 18, most frequently used for diagnosis based on the Positron Emission Tomography method. A radio-labelled glucose analogue that allows glucose-consuming cells such as tumoral cells to be displayed.

Free radical: an extremely reactive chemical entity which contains a redundant electron and which is at the origin of later chemical transformations.

Galenic: study of the method of administering a medication.

Gamma (γ): radiation of a shorter wavelength than X-rays emitted by certain radionuclides and with very high energy, usable for diagnosis imaging.

Generator: tool for the production of a radioisotope by the decay of a parent radionuclide from which it is regularly separated by a physical means (column filtration, extraction).

Gray (Gy): unit of absorbed dose corresponding to one joule per kilogram. The former unit of absorbed dose is the rad, with one gray equalling 100 rads.

Half-life: radioactive half-life, *see Radioactive half-life*. The term biological half-life is also used, which corresponds to the time at the end of which half the quantity of a substance has disappeared or been eliminated from a cell by a biological process.

Internal radiotherapy: method of therapy by irradiation using a sealed radioactive source inserted into a natural cavity or implanted, temporarily or permanently, into the tissues. Synonym for *Curietherapy*. Domain of the radiotherapist.

Intracavity radiation: the emission of rays from a source placed inside a cavity: uterus, throat.

Intraoperative radiation: irradiation during a surgical operation.

Ionising (radiation): electromagnetic or corpuscular radiation capable of producing ions (positively or negatively charged atoms or molecules) directly or indirectly during its passage through the matter. This transformation of the molecules is considered as being destructive and induces a biological change. X and γ rays are considered as weak ionizers compared with β⁻ and above all with α.

Irradiation: exposure to radiation, without physical contact with the radioactive material, not to be confused with contamination in which there is a transfer of radioactive material. Once outside the radioactive field, the person is no longer exposed to the effects of the radiation.

Isotope: all the atoms, the nuclei of which have the same number of protons, form a chemical element. Natural elements amount 92, to which 17 artificial elements must be added. When a given number of protons are associated in an atom with different numbers of neutrons, they represent variant chemical elements called isotopes. In most cases only a few forms are stable, the other unstable forms are called radioisotopes or radionuclides. Of the 109 elements currently known, 28 only exist in an unstable, that is to say radioactive, form. This is the case for uranium, plutonium and radium. There are more than 2,000 known radionuclides.

Label: an entity (simple or complex) which due to its radiation or its colour can be monitored in a complex biological system.

Labelling: method of chemical fixation of a radioisotope on a non-radioactive molecule.

Ligand: molecule joined to the central atom and which confers particular properties on the whole such as better solubility or a better ability to be absorbed by cells.

Metabolic radiotherapy: method of therapy by selective irradiation of a target zone by a molecule participating in the metabolism and labelled with a radioisotope, injected into the patient. Domain of the nuclear physician.

MRI: Magnetic Resonance Imaging, another name for Medical Nuclear Magnetic Resonance.

Neutron: a neutral elementary particle (with no electrical charge), a constituent of the nucleus of the atom together with the proton.

Neutron therapy: external radiotherapy using a neutron beam.

Nocebo: an undesirable effect of a treatment not linked to an active substance, but essentially to an unconscious non-controlled perception on the part of the patient. Opposite of the placebo effect.

Nuclide: atomic nucleus.

Oncology (or cancerology): medical science covering the field of the prevention, detection and treatment of cancers.

PET: Positron Emission Tomography.

Placebo: substance presented in the form of a medication, which contains no active element and which may still have a therapeutic effect on the patient. By extension, the placebo effect stands for any positive effect on the evolution of an illness not linked to an active substance. The opposite, a negative effect, is called nocebo.

Positron: *see Beta-plus.*

Posology: dosage and procedures for administering a medicine.

Proton: a positively charged elementary particle, constituting the nucleus of the atom together with the neutron.

Protontherapy: method of external radiotherapy using a proton beam.

Rad: *see Gray.*

Radiation: a beam of invisible particles or waves emitted by a source. Also, the process of transmission of energy in corpuscular (α, β, etc. particles, etc.) or electromagnetic form (visible light, ultraviolet, infrared, X, γ, etc.)

Radioactive half-life: time at the end of which half the atoms initially present in a radioactive element have disappeared through spontaneous transformation. This period (half-life) varies from one radionuclide to another, but is a precise physical constant for a given radioisotope and is neither influenced by temperature or pressure.

Radioactivity: property of certain radionuclides, which emit particles spontaneously (electrons, protons, neutrons, nuclei) and/or γ or X-rays.

Radiochemical: a radioactive substance not-intended for human use.

Radiochemist: a chemist specialising in the manufacture of radioactive substances, therefore in the nuclear medicine field, a specialist in the development of labelling, and in the nuclear physics field, a specialist in the chemistry of radionuclides.

Radiochemistry: chemistry of substances incorporating a radioactive element.

Radioelement: an element where all the isotopes are radioactive, like for example those of the Plutonium or Uranium groups (term often used wrongly in the place of radionuclide or radioisotope).

Radioisotope: An unstable isotope that decays over the course of time, emitting radiation (*see radionuclide*).

Radiologist: specialist in X-ray imaging.

Radionuclide: radioactive atomic nucleus. Two radionuclides compared with each other are called radioisotopes if they belong to the same family of atoms (*e.g.* the radioisotopes of iodine such as Iodine 123, 124 or 131), and radionuclides in the other cases. The word in the plural "radioisotopes" is frequently used wrongly to designate all radionuclides.

Radiopharmaceutical: radioactive medication intended for diagnosis or therapy in the field of nuclear medicine.

Radiopharmacist: a hospital pharmacist specialising in the labelling and handling of radiopharmaceutical preparations intended for administration to a patient.

Radiopharmacy: a laboratory, principally located in a hospital, equipped to handle radioactive substances for the injection into patients.

Radiophysician: physician specialising in the handling and production of radionuclides.

Radiotherapist: physician specialising in treatment by external radiotherapy. In case the radioactive substance has to be injected into the patient, responsibility is entrusted to the nuclear physician.

Radiotherapy: method of therapy (treatment of a disease) based on the use of radiation, of whatever sort (X-rays, alpha, beta, neutrons, etc.)

Rem: *see Sievert*.

Scanner: an imaging tool using X- or gamma rays that provide virtual sectioning (scans) of the area being analysed.

Scintigraphy: method of imaging based on recording γ radiation emitted by a substance injected into the patient and which concentrates in a particular organ or tissue (heart, thyroid, bones, etc.)

Sealed source: a radioactive substance placed in a sealed container, irradiating but not contaminating. The implants used in internal radiotherapy are sealed sources.

Side effects or **undesirable effects:** disturbance to the state of health of any sort not linked to the principle illness, but more often to the treatment itself. For example: headaches, gastric trouble, nausea and vomiting, hair loss, changes in blood count, as a cause of, or parallel to, chemotherapy.

Sievert (Sv): unit of equivalent dose, corresponding to a corrected dose of the ionising effect of the radiation (for X, β and γ radiation, 1 Sv = 1 Gy). Previously the rem was used, with one sievert equivalent to 100 rems.

Source: origin of radiation. By extension, the radioactive substance itself.

Specific activity: value corresponding to the relationship between the activity of the radionuclide and the total mass of the element present. When the radioisotope is present in its pure form, even in an extremely dilute solution, we talk about a **carrier free** radioisotope solution. It is expressed in becquerels per mass unit.

Specific concentration: value that determines the degree of radioactive substance per volume unit. It is expressed in becquerels per volume unit.

Specific – specificity: describes molecules that only target a single type of cell or receptor.

SPECT: Single Photon Emission Computed Tomography.

Targeted or vectorised radiotherapy: a more general term including metabolic radiotherapy. This nuclear medicine technique consists of treating a specific tissue of the organism with ionising radiation, itself originating from the concentration of a substance which participates in a biological mechanism (the vector), and to which a suitable radionuclide is grafted.

Tomography: radiography providing a clear image of a single cross-section.

US: ultrasound.

Vector: chemical substance which has the property of being recognized by certain macromolecules present in tissue (receptor, enzyme, etc.) and on which is grafted another toxic or radioactive substance, for therapeutic or diagnosis purposes.

X-rays: invisible, short wave, light radiation, produced by a radioactive substance and capable of traversing material.

For Further Reading

In the world, very few nuclear medicine books are written in a language that can be understood by almost everyone, and none of them have been recently updated. The following is a reference list of books generally intended for physicians. They are a good source of complementary information for specific topics, but the required scientific knowledge is of a high level.

In English

Mettler F.A., Guiberteau M.J., *Essentials of Nuclear Medicine Imaging*, Saunders, Albuquerque NM (2006).

Murray I.P.C., Ell P., *Nuclear Medicine in Clinical Diagnosis and Treatment, Vol. 1 and 2*, 2nd edition, Churchill Livingstone, Edinburgh (1998).

Palmer E.L., Scott J.A., Strauss H.W., *Practical Nuclear Medicine*, WB Saunders Company, Philadelphia PA (1992).

Taylor M.D., Schuster M.D., Alazraki N., *A Clinician's Guide to Nuclear Medicine*, 2nd edition, SNM Reston VA (2006).

Wagner H.N., *A Personal History of Nuclear Medicine*, Springer, Berlin (2006).

In French

Blanc D., *La Chimie nucléaire*, Presses Universitaires de France, collection "Que sais-je?" n° 2304 (1987).

Comet M., Vidal M., *Radiopharmaceutiques.Chimie des radiotraceurs et applications biologiques*, Presses Universitaires de Grenoble (1998).

Lambert G., *Une radioactivité de tous les diables. Bienfaits et menaces d'un phénomène naturel... dénaturé*, EDP Sciences, collection "Bulles de sciences" (2004).

Najean Y., *Médecine nucléaire*, Ellipses (1990).

Radvanyi P., *Les Rayonnements nucléaires*, Presses Universitaires de France, collection "Que sais-je?" n° 844 (1995).

Rubinstein M., Laurent E., Stegen M., *Médecine nucléaire. Manuel pratique*, De Boeck Université (2000).

Tubiana M., *Le Cancer*, Presses Universitaires de France, collection "Que sais-je?" n° 11 (2003).

Tubiana M., Dautrey R., *La Radioactivité et ses applications*, Presses Universitaires de France, collection "Que sais-je?" n° 33 (1997).

Tubiana M., Lallemand J., *Radiobiologie et radioprotection*, Presses Universitaires de France, collection "Que sais-je?" n° 2439 (2002).

In German

Elser H., *Leitfaden Nuklearmedizin*, 2nd edition, Steinkopf Darmstadt, Springer (2003).

Pabst H.W., Adam W.E., Hör G., Kriegel K., Schwaiger M., *Handbuch der Nuklearmedizin*, Gustav Fischer, Stuttgart, New York (1996).

Schicha H., Schober O., *Nuklearmedizin, Basiswissen und Klinische Anwendung*, Schattauer Germany (2003).

Contents

Preface 5

Introduction and Definitions 7

Chapter I. Nuclear Medicine, what for? 9
I. The Case of Thyroid Cancer 11
II. The Diagnosis Aspect 12
III. The Therapeutic Aspect 17
    1. Cancer Therapy 17
    2. Another Therapeutic Application: Rheumatology 22
IV. Other Aspects in this Area 22

Chapter II. A Little Bit of History 25

Chapter III. Some Basic Notions of Radiation 31
I. Different Types of Radiation 33
II. Measurement Units and Doses 38
III. Radionuclides for Nuclear Medicine 45
    1. Gamma Emitters (γ) 46
    2. Positron Emitters (β^+) 47
    3. Electron Emitters (β^-) 48
    4. Alpha Emitters (α) 49
    5. Radionuclides for Brachytherapy and External Radiotherapy 49
    6. Other Radionuclides 50
Summary 51

Chapter IV. Gamma Ray Imaging 53
I. Nuclear Medicine Imaging Methods 58
   1. Scintigraphy 60
   2. The Products used in Scintigraphy 62
II. Imaging Tools 64
III. Detection of the Sentinel Node 66
Summary.. 68

Chapter V. PET Imaging: Positron Emission Tomography 71
I. The Imaging Principle 73
II. The Radiation Source 74
III. The Labelled Product: Fludeoxyglucose 75
IV. Production and Equipment 77
V. Applications in Cancerology 78
VI. Applications beyond Oncology 79
VII. Positron Emitters Evolution.. 80
Summary.. 81

Chapter VI. Therapeutic Methods 83
I. Metabolic Radiotherapy.. 84
II. Local Radiotherapy.. 86
III. Radioimmunotherapy 87
IV. Targeted Radiotherapy.. 93
V. Alphatherapy and Alpha-immunotherapy 94
VI. Neutron Capture Therapy.. 99
VII. Radiotherapeutic Substances.. 102
VIII. The Dose Issue 103
IX. Mechanism of Action – The Bystander Effect 104
X. The Limitations.. 106
Summary.. 108

Chapter VII. The Development of Radiopharmaceuticals 109
I. The Molecule Discovery Phase 111
II. Pharmacological and Preclinical Studies 111
III. Pharmacokinetics 112
IV. Toxicological Analysis 113
V. Phase I Clinical Studies.. 116
VI. Phase II Clinical Studies 117
VII. Phase III Clinical Studies.. 120
VIII. Regulatory Issues and Registration 123

IX. Marketing 125
X. Post-marketing Authorisation and Drug Monitoring................. 125
Summary.. 127

Chapter VIII. The Production of Radiopharmaceuticals..................... 129
I. Definitions... 129
II. The Principles behind Radionuclides Production 131
 1. Particle Accelerators 131
 2. Generators 133
 3. Reactors.. 134
 4. Fission Products... 134
III. The Production of Vectors and Ligands 134
IV. The Industrial Production of Radiopharmaceuticals 135
V. Transport and Logistics 137
VI. Radiopharmacies... 138
VII. Nuclear Medicine Centres in the World 140
Summary.. 145

Chapter IX. Future Prospects 147
I. Hybrid Imaging Tools.. 147
II. Individualised Medication 148
III. Orphan Diseases and Orphan Drugs 151
IV. Ethical and Regulatory Limitations 153
 1. Regulation and Administration..................................... 153
 2. Side Effects and Toxicity 154
 3. Dosage and Indications Extensions 155
V. Politics and Legislation 156
VI. The Future 158

Glossary.. 161

For Further Reading 169

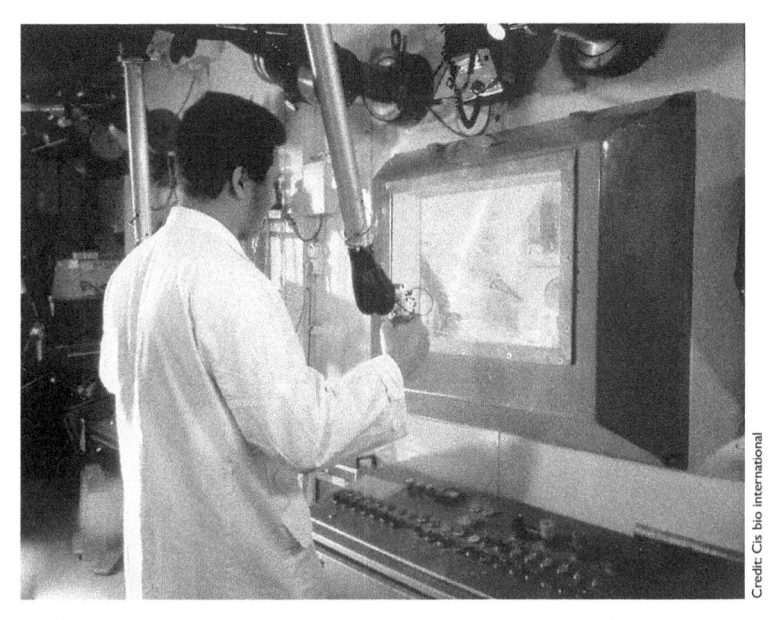

Radiopharmaceutical preparation via remote control through thick leaded glass.

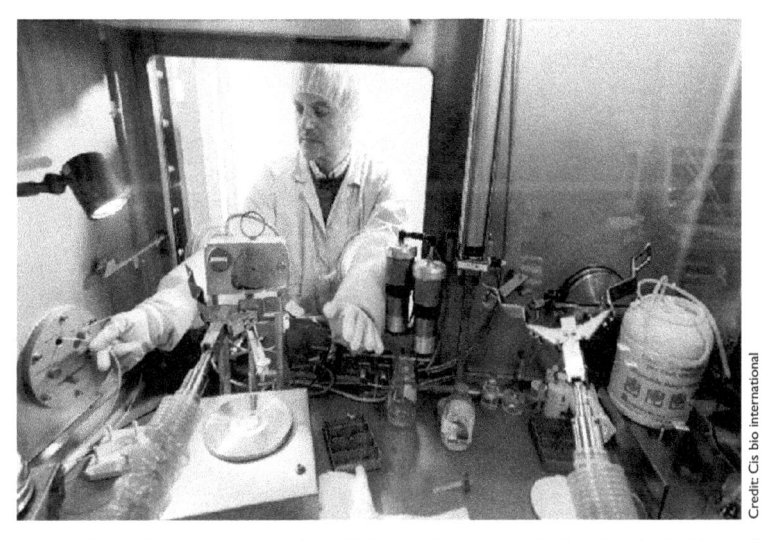

Preparation from the rear opening of an FDG manufacturing cell, showing the inside tools.

30 MeV cyclotron for the manufacturing of radionuclides (e.g. Thallium 201, Iodine 123, Indium 111) *(IBA equipment)*.

18 MeV cyclotron for the production of Fluorine 18. This view of the open cyclotron shows the quadrupoles which enable proton acceleration *(General Electric equipment)*.

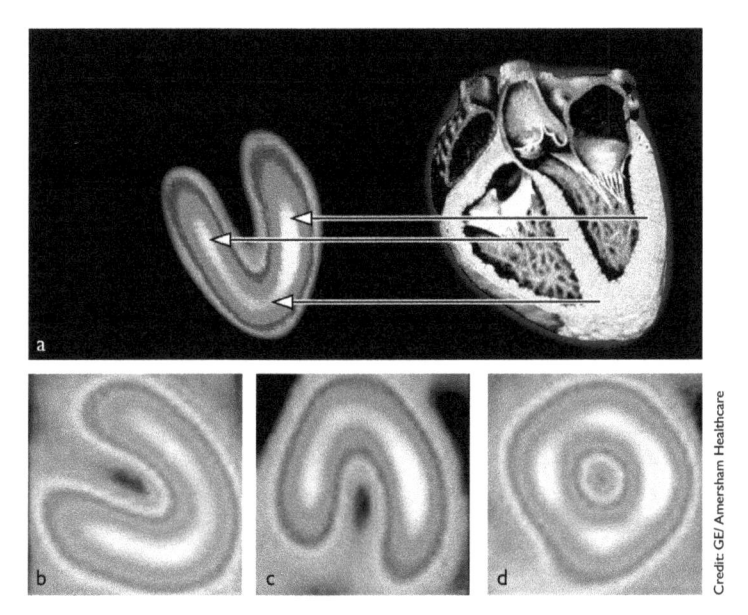

(a) Cardiac image obtained after injection of Technetium 99m tetrofosmin. The radioactive substance is integrated into the active healthy cells. The cells' incorporation level is directly linked to the rate of radioactivity, which is represented by a scale of colours depending on the level of intensity. One can therefore clearly superimpose the muscular part of the heart that will show a U-shape in a section scanned from the front (b), or from the side (c) and a doughnut shape (d) if seen from the top. Any reduction in size of this area, any missing colours indicate an ischaemic area. If the image of this area does not resume to a normal colour set following a stress test (bicycle or treadmill), one can consider that this area is necrosed. Medical treatment will be adapted accordingly.

Pulmonary ventilation imaging is obtained by inhalation of a radioactive gas. The higher the pulmonary capacity is, the higher the radioactive area will be. Colours with varying intensities indicate the radioactive substance concentration level. Obviously, the image showing a badly ventilated lung (right) is incomplete compared to that of a normal lung (left).

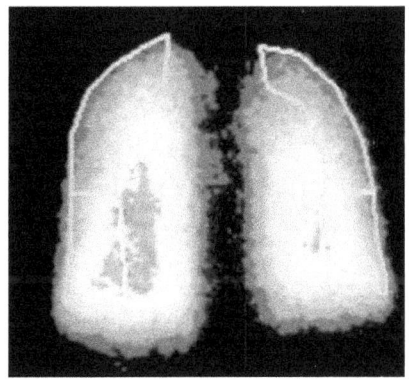

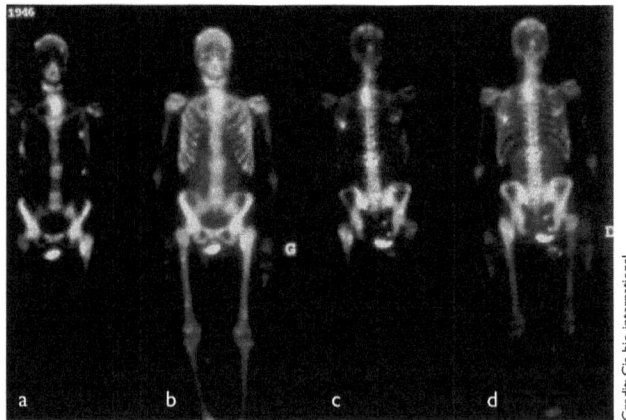

Specific drugs, such as MDP and EDTMP, allow for easy visualisation of cancer patient bone metastases. In order to demonstrate the efficacy of these drugs in therapeutic use, it is possible to firstly inject the molecule labelled with an imaging radionuclide such as Technetium 99m (images a and c correspond to both anterior and posterior views of a same patient). The same product substituted with Samarium 153 shows that it binds on the same tumours and metastases (images b and d). This comparison is exceptionally possible as Samarium is both a gamma emitter (imaging) and a beta-minus emitter (therapy). Theoretically, this radionuclide is used for therapy only, and will only have a destructive effect in the highlighted areas. The low background noise on the arms and legs demonstrates that the binding takes place on the tumoral cells only, and not on the bones themselves.

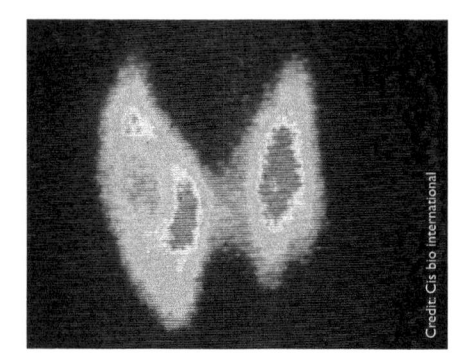

Injections of Iodine 131 show a very clear and specific image of the thyroid, as the latter absorbs iodine very easily. The slightly emitting area located on the left side of the image (right thyroid lobe) clearly indicates the presence of an isolated and well-delimited cold nodule. The subsequent ultrasound analysis will help define whether the nodule is a solid benign tumour (cellular excrescence that prevents the organ from functioning properly) or a cyst. In the case of a hot or hyper-binding nodule, Iodine 131 (gamma and beta-minus emitter) can be used at low doses for imaging and at higher doses for therapy purposes via the destruction of the cells on which it binds.

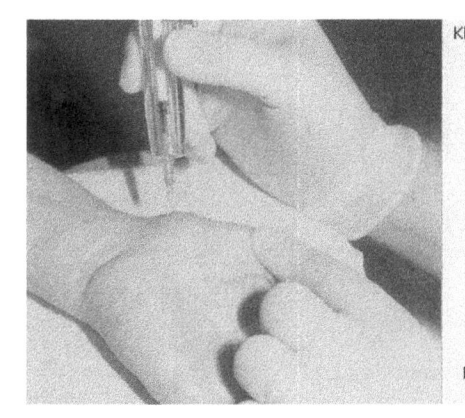

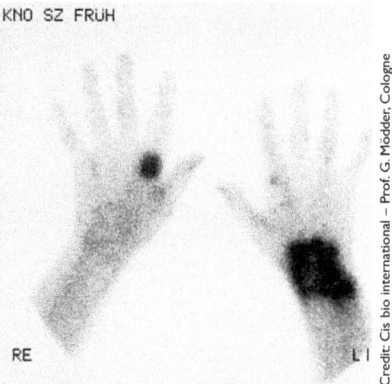

The image on the right shows the inflamed hand areas (finger on the left hand and wrist on the right hand), and is obtained after injection of Technetium 99m. The image on the left shows the treatment procedure for polyarthritis; the dark areas completely disappear a few weeks after a single localized synoviorthesis course of treatment.

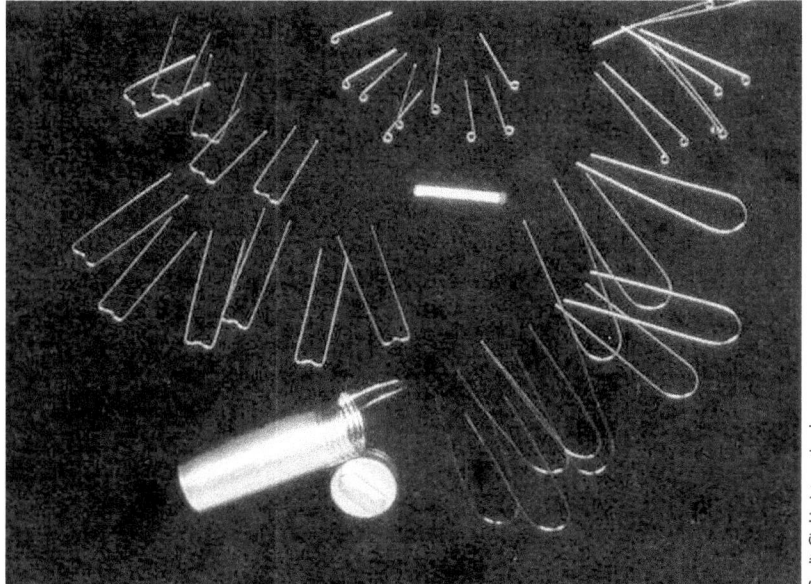

Examples of brachytherapy implants made of metallic wires to be introduced directly into the tumour for a precise length of time, during which the surrounding cells are irradiated and destroyed. The match in the middle of the picture gives a good idea of the size of these implants; it also gives, in an indirect manner, a rough idea of the size of the tumours in which they will be placed. The physician will adapt the length of the wire to the actual size of the area to be treated.

V

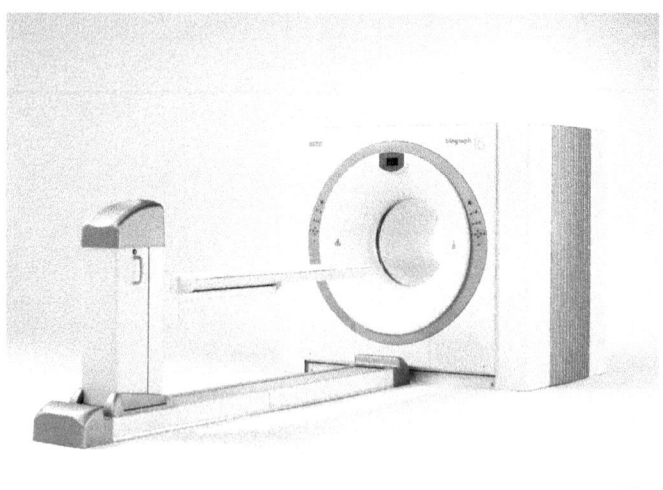

Credit: Siemens

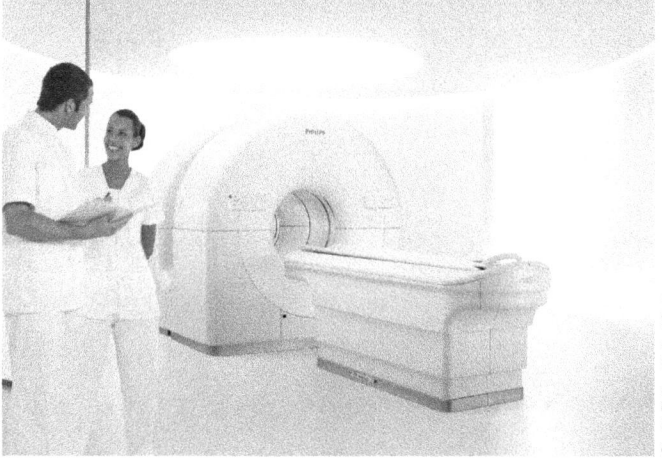

Credit: Philips Medical Systems

For a neophyte, nuclear medicine equipment does not seem to differ much from MRI equipment. Both scanners as shown on this page are mixed PET-CT cameras. The patient is injected a dose of FDG, then is laid out on the examination table and brought twice through the scanners successively for the morphological analysis (X-ray of all organs) and functional imaging (measuring of radioactivity levels in specific tissues). The superimposition of both images pinpoints with great accuracy the cells involved in the disease (tumours, metastases, but also inflammation and infection sites), thus providing very precise information on the disease's development stage.

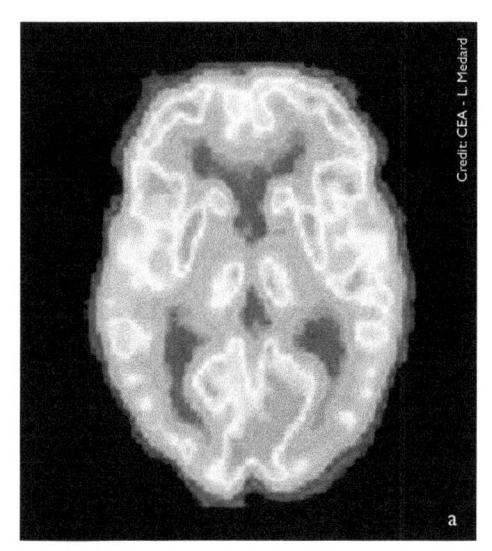

Positron Emission Tomography (PET) is an ideal tool for studying brain behaviour (bottom picture). As a consequence of the rapid distribution of the Carbon 11 or Fluorine 18 labelled molecules in the brain and of the short half-life of these radionuclides, it is possible to observe how flawed brain cells compare with healthy ones, as well as to study how the activated brain areas respond to specific stimuli (think, look, move, etc...). The top image on the left shows a PET scan of a healthy subject's brain.

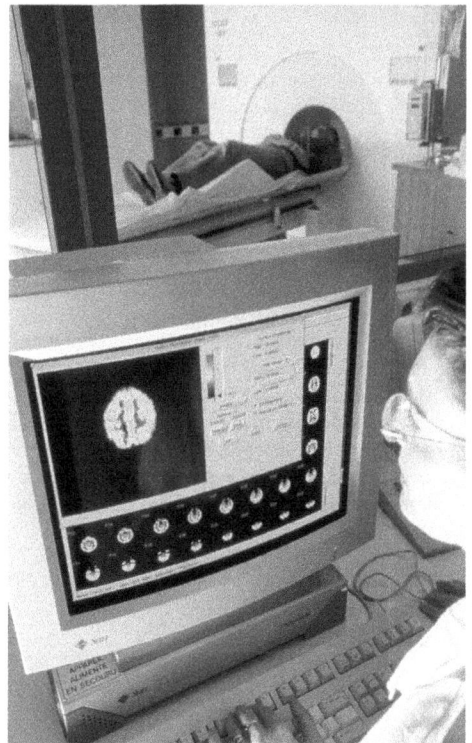

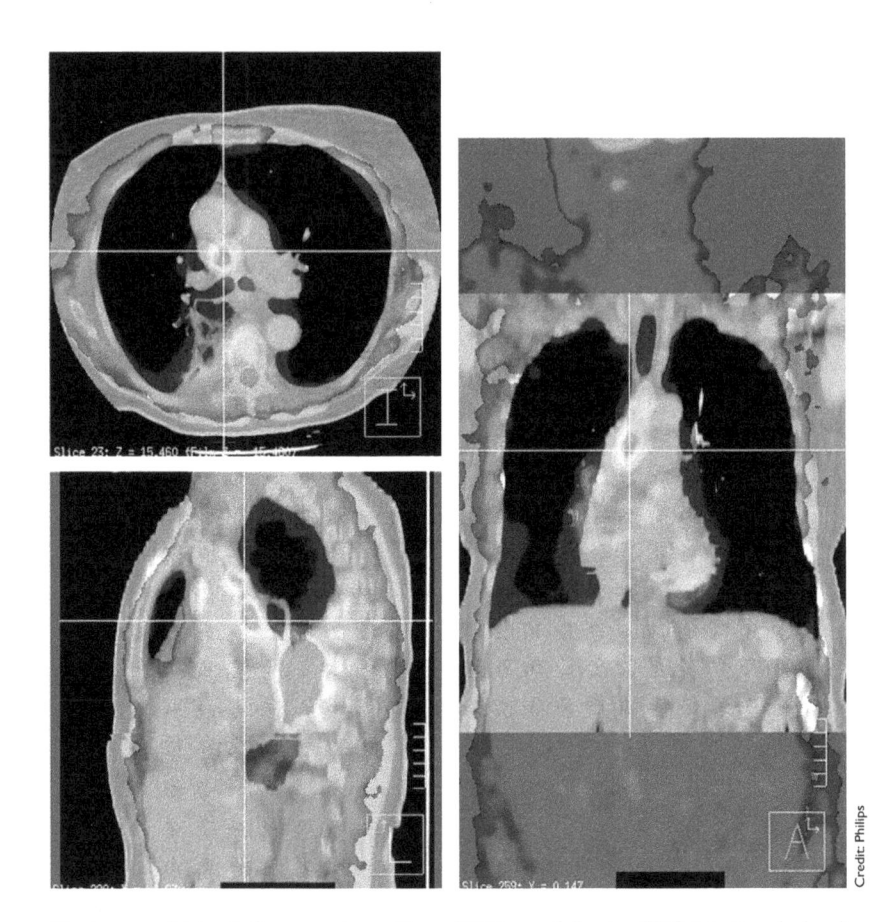

The images obtained after superimposition of PET scan following FDG injection (colour) with X-ray scan (gray background) with a PET-CT camera clearly show the three-dimensional positioning of tumours and metastases. The colours are relative to a scale that is linked to the intensity of the radiation, and thus indirectly to the amount of fluorinated glucose locally absorbed. For example, red colour corresponds to a strong uptake of glucose while blue colour remains close to the background noise. It is important to mention that, beside the tumoral cells, the heart (large red spot) and the brain also consume high levels of glucose.